£35.00

WARSASH MARITIME SCHOOL
PART OF SOLENT UNIVERSITY

GLOBAL NAVIGATION
A GPS USER'S GUIDE

Neil Ackroyd and Robert Lorimer

|L|L|P|

LONDON   NEW YORK   HAMBURG   HONG KONG
LLOYD'S OF LONDON PRESS LTD.
1990

Lloyd's of London Press Ltd.
One Singer Street,
London EC2A 4LQ
Great Britain

USA AND CANADA
Lloyd's of London Press Inc.
Suite 523, 308 Broadway
New York, NY 10012 USA

GERMANY
Lloyd's of London Press GmbH
59 Ehrenbergstrasse
2000 Hamburg 50

SOUTH-EAST ASIA
Lloyd's of London Press (Far East) Ltd.
Room 1101, Hollywood Centre
Hollywood Road
Hong Kong

© Neil Ackroyd and Robert Lorimer
1990

First published 1990

All rights reserved. No part of this publication may be reproduced, stored in a retrieval system, or transmitted, in any form or by any means, electronic, mechanical, photocopying, recording or otherwise, without the prior permission of Lloyd's of London Press Ltd.

British Library Cataloguing in Publication Data
Ackroyd, Neil
Global navigation.
1. Satellite navigation equipment
I. Title II. Lorimer, Robert
629.045

ISBN 1–85044–232–0

Text typeset in 9pt on 11pt Century Schoolbook by
Megaron, Cardiff, Wales
Printed in Great Britain by
Bookcraft Ltd., Midsomer Norton

To our wives, Kathryn and Christine

Acknowledgements

We would like to thank our companies Racal Survey Ltd and Geografix Ltd for their cooperation and support during this project. We would also like to thank the INMARSAT Organisation for the provision of relevant materials

NEIL ACKROYD, ROBERT LORIMER

Disclaimer

The authors would like to state that the opinions expressed in this book are their own and do not necessarily represent those of their companies or the directors thereof.

Contents

	Page
Acknowledgements	iv
Disclaimer	iv
List of Illustrations	xi
List of Tables	xiv
List of Abbreviations	xv
Bibliography	xviii

CHAPTER 1	SATELLITES AND NAVIGATION	1
1	Global positioning	1
	Introduction	1
	1.1 The global positioning concept	2
	1.2 The navigator's choice	4
	1.2.1 Regional navigation systems	4
	Omega	6
	Loran C and Chayka	6
	Decca	9
	1.2.2 The satellite alternatives	9
	Star-Fix	10
	Geostar	12
	Proposed systems	12
	NAVSAT	13
2	The global positioning systems	14
	2.1 The first satellite systems	14
	Transit	14
	2.2 Satnav and GPS: a perspective	16
	2.3 Navstar GPS—a system description	17
	2.3.1 The system design	17
	2.3.2 Current configuration	18
	2.3.3 Navigating with GPS	19
	Pseudo-range measurements	20
	The calculation of position	23
	2.3.4 GPS system accuracies	23
	Single receiver GPS accuracies	25
	Enhanced accuracy levels	25

vi   *Contents*

CHAPTER 2   GPS AND THE SHIP — 27

Introduction — 27
1. GPS and the electronic bridge — 27
   1.1 The integrated ship — 27
   1.2 GPS and the operator — 31

2 The GPS receiver — 31
   2.1 The GPS receiver types — 31
       Parallel multi-channel receivers — 31
       Slow and fast sequencing receivers — 32
       Multiplexing receivers — 33
   2.2 The hybrid GPS receiver — 34
   2.3 A practical guide to purchase — 36
       2.3.1 Navigation features — 37
       2.3.2 Operating modes — 37
       2.3.3 User interface — 38
       2.3.4 Hardware interfacing — 39
       2.3.5 Hardware integration — 40
       2.3.6 Power supplies — 41
       2.3.7 Antenna installations and cabling — 41
       2.3.8 Differential features — 42
       2.3.9 Service agreements and software upgrades — 43

CHAPTER 3   DIFFERENTIAL GPS — 45

1 The differential concept — 45
   1.1 Introduction — 45
   1.2 System design — 46
   1.3 Techniques of correction — 49
       Advantages and disadvantages — 49
   1.4 Pseudolites — 52
   1.5 Differential GPS and selective availability — 53
   1.6 The Radio Technical Committee for Marine Services — 55
       The RTCM message format — 55
       Data rate — 57
   1.7 GPS integrity monitoring and quality control — 58

2 Data links — 59
   Introduction — 59
   2.1 The compromise — 60
   2.2 The differential options — 61
       Long wave/low frequency (30 kHz to 300 kHz) — 61
       Medium frequency (300 kHz to 3 MHz) — 64
       High frequency (3 MHz to 25 MHz) — 66
       Very high frequency and ultra high frequency (30 MHz to 300 MHz) — 66
       The satellite dimension — 66

Contents vii

CHAPTER 4 GPS: APPLICATIONS AND IMPLICATIONS	69
Introduction	69
1 GPS and coastal navigation	69
Coastal navigation, a definition	69
Physical factors	70
Technological factors	70
1.1 Conventional electronic positioning services	71
1.1.1 User access	71
1.1.2 User cost	73
1.1.3 Operator costs	74
1.1.4 Accountability	74
1.2 Traffic management and separation schemes	74
1.2.1 The specification of traffic separation schemes	76
1.3 Conventional aids to navigation	77
1.4 Navigation equipment on merchant vessels	80
1.5 Alternative GPS applications	83
1.5.1 GPS for merchantmen	83
1.5.2 GPS for fishermen	84
1.5.3 GPS for oilmen	87
1.5.4 GPS for coastguards and policemen	87
Vessel surveillance	88
Vessel identification	88
Vessel interception	88
1.5.5 GPS for yachtsmen	89
1.5.6 GPS for EEZ management	90
2 GPS for port positioning	90
Introduction	90
2.1 Conventional port positioning services	91
2.1.1 Hydrographic surveys	91
Low technology surveys	91
Semi-automated surveys	92
Fully automated surveys	95
Software complexity	95
Software language	95
Computer hardware	96
2.1.2 Dredging operations	96
2.1.3 Buoy movements and monitoring	96
2.1.4 Vessel navigation and pilotage	98
Physical specification	99
Operating specifications	99
Software specification	99
Display specification	99
2.2 The GPS port positioning service	101
2.2.1 Compatibility	101
Co-ordinate systems	101
Communication protocols	101

				Physical characteristics	103
		2.2.2	Levels of service		103
			The precision service		103
			The accurate service		104
			(1)	Geography	105
				Conventional	105
				GPS	105
			(2)	Operational considerations	106
				Conventional	106
				GPS	106
			(3)	Demand Restrictions	107
				Conventional	107
				GPS	107
			(4)	Reliability and availability	107
				Conventional	107
				GPS	107
				The standard service	107

3 Position and data reporting — 108
  Introduction — 108
  3.1 INMARSAT's Standard C satellite communications service — 109
    The enhanced group call facility — 109
    INMARSAT position reporting and surveillance service — 112
    3.1.1 Scenario one: A basic position reporting system — 112
    3.1.2 Scenario two: A position reporting and display system — 112
      (1) Communications — 114
      (2) Position display and logging — 114
      (3) Position report database — 114
      (4) Reports and data export — 114
    3.1.3 Scenario three: An electronic fleet management system — 115
  3.2 Integrated fleet management — 115

CHAPTER 5   THE GPS DETAIL — 119

Introduction — 119
1 The system design and implementation — 119
  Introduction — 119
  1.1 The space segment — 120
    Orbit design — 121
      Navstar GPS — 121
      Glonass — 122
    The satellites — 122
      Navstar GPS — 122
      Glonass — 127
  1.2 The ground control segment — 127
    Civilian GPS information centre — 129
    Glonass — 129

			Page
	1.3	The user segment	131
		The GPS receiver design	131
		From noise to signal	132
		From signal to numbers	133
		From numbers to code	133
		From numbers to phase	135
		From measurement to position	136
	1.4	The system status	136
2	GPS: The signals	138
	Introduction	138
	2.1	The GPS clock	138
		System time	140
		Einstein and GPS	141
	2.2	The GPS frequencies	141
		2.2.1 The GPS carriers	141
		Navstar	141
		Glonass	143
		2.2.2 The codes and spread spectrum	143
		Navstar	144
		2.2.3 The C/A code and the range	145
		2.2.4 The P code and how	147
		2.2.5 The Y code and why?	148
	2.3	Codeless GPS	148
		2.3.1 carrier aided filtering	149
		2.3.2 Phase differencing	150
3	Pseudo-ranging for position	152
	Introduction	152
	3.1	The pseudo-range	152
	3.2	The satellite's position	152
		Kepler and GPS	154
	3.3	The computation to position	154
	3.4	Position aiding	156
		3.4.1 Height aiding	156
		3.4.2 Clock aiding	159
	3.5	The position reference frame	159
		3.5.1 The ellipsoid	160
		3.5.2 Cartesian reference frames and spheroids	161
		3.5.3 The WGS 84 spheroid	162
4	GPS positioning quality	163
	Introduction	163
	4.1	Real-time quality control	163
		4.1.1 The dilution of precision	164
		The various DOPs	166
		4.1.2 DOPs and UEREs	166

x *Contents*

4.2 The completed constellation	167
4.3 Confidence levels	168

CHAPTER 6 STANDARD C DETAIL: INMARSAT AND GLOBAL SATELLITE COMMUNICATIONS — 171

1 History in brief	171
2 The INMARSAT organization	171
3 The INMARSAT Standard A service	174
3.1 The Standard A ship earth station (SES)	175
3.2 The Standard A coast earth station (CES)	177
3.3 INMARSAT's current (first generation) space segment	178
Frequency allocation	181
3.4 INMARSAT second generation space segment	182
3.5 INMARSAT third generation space segment	184
4 The INMARSAT Standard C system	184
4.1 The Standard C ship earth station	185
4.2 The Standard C coast earth stations	187
4.3 The Standard C space segment	188
4.4 The network co-ordination station (NCS) common channel	188
4.5 The coast earth station (CES) TDM channel	189
4.6 Ship earth station (SES) signalling channel	189
4.7 SES messaging channel	190
4.8 Inter-station channels	190
Inter-region channels	190
5 Standard C services	190
5.1 Store and forward data and messaging	191
5.2 Shore-to-ship message transfer	191
5.3 Ship-to-shore message transfer	192
5.4 One-way position and data reporting	193
5.5 Report format	194
5.6 Polling	195
5.7 Enhanced group calls	195
5.8 Standard C and the global maritime distress and safety system	198
INDEX	199

List of Illustrations

Fig.		Page
1.	The global ship	2
2.	A GPS satellite	3
3.	Omega station locations	5
4.	Loran C and Chayka coverage	7
5.	Decca coverage	8
6.	Star-Fix system coverage	10
7.	Geostar: existing and planned coverage	11
8.	Three out of seven planes in a GEO/HIO mix	13
9.	Transit orbit arrangement	14
10.	Transit system principle	15
11.	GPS orbit arrangement	17
12.	English Channel GPS coverage as at March 1990	18
13.	P code and C/A code modulations	19
14.	The pseudo-range	21
15.	GPS and position aiding	22
16.	Geometry effects	24
17.	A differential GPS scatter plot	26
18.	Artist's impression of the integrated bridge, including 2690 BT ARPA, CVP 3500 and MNS 2000 with LSR 4000 screen	28
19.	Typical display of LSR 4000 Live Situation Report	29
20.	The Racal Decca MNS 2000 on board Sealink ferry *Hengist*	30
21.	A dual channel fast scanning receiver	33
22.	Shipmate RS 5310—a multiplexing receiver	34
23.	MNS 2000G display	35
24.	An integrated GPS/Loran C receiver	36
25.	A generic receiver menu design	39
26.	Shipmate RS2500 colour plotter	41
27.	Magnavox MX 4400 GPS positioning and navigation system	43
28.	Differential GPS	46
29.	Common view (block shift) correction technique	48
30.	Pseudo-range error sources	50
31.	Pseudo-range comparisons	51
32.	Selective availability plot	54
33.	Type 1 message format	56
34.	100 kHz transmitter	62
35.	Existing differential coverage provided to the oil exploration industry	63

xii *Illustrations*

36.	A medium frequency differential GPS station	65
37/38.	Diffcell—a combined GPS receiver and VHF transmitter	67
39.	Coastal DGPS service	70
40.	Precise positioning in the North Sea better than 10m	72
41.	NavGraphic II—and ECDIS using published nautical charts	76
42.	Co-sited MF DGPS station and lighthouse	79
43.	2690 BT ARPA on board Sealink ferry *Horsa*	82
44.	The electronic fishing vessel: wheelhouse of crabber William Henry II	85
45.	A seismic vessel towing a sound source	86
46.	SARSAT-COSPAS	89
47.	Semi-automated survey package	93
48.	Automated survey system	94
49.	Port dredging system	97
50.	Port pilotage system	100
51.	Port DGPS service	102
52.	A differential system	104
53.	Micro-fix beacon	106
54.	SafetyNet call Navarea II	110
55.	FleetNet call	111
56.	Position reporting service	113
57.	The fleet network	116
58.	Satellite positioning system configuration	120
59.	Satellite constellation Bird-cage	121
60.	A Navstar GPS satellite	123
61.	Navstar GPS Block 1 in-plane constellation	124
62.	Block 1 and planned Block 2 positions	126
63.	Operational ground control system Navstar GPS	128
64.	Navstar GPS CGIC interface	130
65.	GPS receiver in situ!	131
66.	GPS signal spectrum	132
67.	Receiver design	134
68.	Current and projected launch build-up	137
69.	Oscillator performance comparison	139
70.	The GPS clock	140
71.	The GPS carrier frequencies	142
72.	Binary biphase modulation	144
73.	GPS signal 'in quadrature'	145
74.	The GPS codes	146
75.	The complex GPS signal structure	149
76.	The improvements of continuously integrated Doppler	150
77.	Phase differencing	151
78.	The pseudo-range	153
79.	GPS ranging principle	155
80.	GPS range to position	156
81.	Height-aiding	157
82.	WGS 84 Geoid-spheroid separation, Euroshelf (North Sea, 5 metre contour)	158

83.	A simple ellipse	161
84.	Cartesian and geodetic frameworks	162
85.	Diamond of errors	164
86.	The dilution of precision	165
87.	Geometry outages (18 + 3 active spares)	168
88.	The GPS accuracies	169
89.	The INMARSAT system	172
90.	INMARSAT global coverage showing 0° and 5° elevation contours	173
91.	A Standard A radome installation (Betsy Ross Ade)	175
92.	Below-decks installation	176
93.	CES parabolic antenna, Fucino, Italy	179
94.	Marecs satelitte	180
95.	Intelsat V satellite	181
96.	Marisat satellite	182
97.	INMARSAT 2 satellite	183
98.	Standard C antenna	185
99.	The Thrane and Thrane terminal	186
100.	Geographic addressing	196
101.	SafetyNet distress alert placed through an RCC SES	197

List of Tables

Table		Page
1.	The dilution of precision	24
2.	Hybrid GPS receivers	35
3.	Differential error budget	47
4.	RTCM message types	57
5.	Equipment rationalization with the introduction of GPS, DGPS and the electronic chart	81
6.	Block 1 Status (December 1989)	122
7.	Navstar numbering schemes	125
8.	Active Glonass satellites	127
9.	Block 2 launch schedule	136
10.	The Keplerian parameters	154
11.	Associated figures of the earth	160
12.	The Standard A coast earth stations (CES)	178
13.	INMARSAT's current (first generation) satellites	179
14.	The Standard C terminal forecast	187
15.	Proposed Standard C coast earth stations	188

List of Abbreviations

AMVER	American Version
AOR	Atlantic Ocean region
ARENTO	The National Telecommunications Organization of Egypt
AUSREP	Australian Report
C/A Code	Clear or Coarse/Acquisition code
CEP	circular error probabilities
CES	coast earth station
CGIC	Civil GPS Information Centre
CID	continuously integrated doppler
CNID	closed network identification number
CRT	Cathode Ray Tube
CTS	Conventional terrestrial system
CSDN	Circuit switched data network
DDOP	differential dilution of precision
DGPS	differential GPS
DOP	Dilution of precision
DOT	Department of Transport
DTE	Data circuit terminating equipment
ECDIS	electronic chart display systems
EGC	enhanced group call
EDOP	Easting dilution of precision
EEZ	Exclusive Economic Zone
EMCS	Engine Monitoring and Control Systems
EMPF	electro-magnetic position fixing system
EPFS	Electronic Position Fixing System
ESA	European Space Agency
ESMS	electronic ship management systems
ETA	estimated time of arrival
FAA	Federal Aviation Authority
GDOP	Geometric dilution of precision
GLA	General Lighthouse Authorities
Glonass	Global Navigation Satellite System
GMDSS	global maritime distress and safety system
GPS	Global Positioning System
HDOP	Horizontal dilution of precision
HOW	hand over word
IEC	International Electro-Technical Commission

List of Abbreviations

IMO	International Maritime Organization
INMARSAT	International Maritime Satellite Communication Organisation
INS	Integrated Navigation Systems
Intelsat	International Satellite Telecommunications Organisation
IOR	Indian Ocean region
ISO	International Organisation for Standardisation
JAPREP	Japanese report
JRC	Japan Radio Corporation
Lanby	Large automated navigation buoy
LCD	Liquid Crystal Displays
LED	Light Emitting Diodes
MAFF	Ministry of Agriculture Fisheries and Food
MCS	master control station
MEM	macro encoded station
MOPS	Minimum operational performance standards
MRB	Marine Radio Beacon
MRB/DF	Marine Radio Beacon/Direction Finder
MTPS	Minimum technical performance standards
NANUS	Navigation Notices to Users
Navstar GPS	Navigation Satellite Timing and Ranging Global Positioning System
NCS	Network Co-ordination Station
NDOP	Northing dilution of precision
OCC	Operation Control Centre
Opscap	Operations Status Capability
PC	personal computer
P code	Precision code
PDOP	Position dilution of precision
POR	Pacific Ocean region
PRN	pseudo-random number
PPS	Precise Positioning Service
PSDN	Packet switched data network
Racon	Radar Beacon
RDF	Radio Direction Finders
RDOP	relative dilution of precision
RDSS	Radio Determination Satellite Systems
RTCM	Radio Technical Committee for Marine Services
SA	Selective Availability
SAR	Search and Rescue
SatComms	Satellite communications
SEP	spherical error probabilities
SES	ship earth station
SOLAS	Safety of Life at Sea
SPS	Standard Positioning Service
SV ID	space vehicle identity
SVn	space vehicle number
TDOP	Time dilution of precision
UDRE	user differential range error

UERE	user equivalent range error
UKCSG	United Kingdom Civil Satellite Group
USCGT	US Coastguard
UTC	Universal Time Coordinated
VDOP	vertical dilution of precision
WGS 84	World Geodetic System 1984

Bibliography

GPS System and Signals

Anodina, T. G. & Prilepin, M. T. — "The GLONASS System"; Proceedings Fifth International Geodetic Symposium on Satellite Positioning, Las Cruces, May 1989.

Anodina, T. G. — "The Glonass System Technical Characteristics and Performance"; International Civil Aviation Organisation, Working Paper Special Committee on Future Air Navigation Systems (FANS), May 1988.

Ashkenazi, V. — "Positioning by Second Generation Satellites GPS and NAVSAT"; Proceedings 2nd Int. Hydro. Conf. 1984.

Ashkenazi, V. — "Precise Static and Dynamic Positioning by GPS"; Proceedings of HYDRO 1986. 16–18 Dec. 1986.

Ashkenazi, V. — "To Code or Not to Code : That is Not the GPS Question"; *Land and Mineral Surveying*, Vol. 5, Feb. 1987. pp. 59–62.

Bell, J. - "Navigation for Everyman and his Bomb"; *New Scientist*, 11 Oct 1984. pp. 36–40

Dale, S. & Daly, P. — "Recent Observations on the Soviet Unions's Glonass Navigation Satellites"; IEEE PLANS 1986 Position and Navigation Symposium, Las Vegas.

Daly, P. — "Aspects of the Soviet Union's Glonass Satellite Navigation System"; *The Journal of Navigation*, Vol. 41, No. 2.

Danchik, R. — "Navy Navigation Satellite System Status"; Position Location & Navigation Symp. Orlando, 22 Nov.–2 Dec. 1988.

Denaro, R. P. — "Navstar GPS Offers Unprecedented Navigational Accuracy"; *Microwave Systems News*, Vol. 14, No. 12, 1984.

Dennis, A. R. — "STARFIX"; Proceedings PLANS 86, Position Location and Navigation Symposium Las Vegas, November 1986.

Geostar — "Understanding Radio Determination Satellite Service"; Geostar Corporation, Washington D.C., May 1989.

Jorgenson, P. S. — "Relativity Correction in GPS User Equipment"; IEEE PLANS 1986, New York 1986.

Matthos, P. — "A Low Cost Hand-Held GPS Receiver"; IEEE Symposium Satellite Communications and Positioning. 1988.

N.A.T.O — "Military Agency For Standardisation: Navstar GPS System Characteristics — Preliminary Draft"; September 1985.

Van Dierendonck *et al.* — "The GPS Navigation Message" GPS Vol. 1 (Red Book), Institute of Navigation 1980, Washington D.C.

Pratt, T. — "The Electronics and Hardware of GPS Receivers"; Seminar on the Global Positioning System, Univ. Nottingham, 12–14 April 1989.
Remondi, B. — "Performing Centimetre-level Surveys in Seconds: Initial Results"; NOAA Technical Memo. NOS NGS 42.
Rockwell Int. Corp. — "Interface Control Document : Navstar GPS", ICD GPS2000, 26 Sept. 1984.
Rosetti & Diederich — "Trends in the Evolution of Global Satellite Navigation Systems"; IEEE Symposium satellite Communications and Positioning, 1988.
Spilker, J. J. Jnr. — "GPS Signal Structure and Performance Characteristics" Papers Published in Navigation (Red Book) Vol. 1, 1980. The Institute of Navigation, Washington D.C.
Wells, D. *et al.* — "A Guide to GPS Positioning"; The Canadian GPS Associates, Fredericton, 1987.

Differential GPS

Ackroyd, N. Daly, P. J. & Roberts — "Offshore Survey Control with Navstar GPS : Our North Sea Experience"; International Geographical Symp. Brisbane, 1988.
Ackroyd, N. & Daly, P. J. — "Experience and Operational Results of Differential Kinematic GPS"; Royal Institute of Navigation NAV 89 Conference, October 1989.
Blackwell, E. G. — "An Overview of Differential GPS Methods" ; Navigation Vol. 32, No. 2, 1985. pp. 114–125.
Blanchard, W. — "Differential GPS"; Seminar On Global Positioning System, Univ. Nottingham, 12–14 April 1989.
Beser and Parkinson — "The Application of Differential GPS in The Civil Community", *Navigation*, Vol. 29, No. 2, 1982.
Enge, P. K. — "Differential Operation of GPS"; *IEEE Communication Magazine*, Vol. 26, No. 7, July 1988, pp. 48–59.
Enge, P. K. — "Marine Radio Beacons for the broadcast of Differential GPS Data"; Proceedings IEEE PLANS, Las Vegas, 1986.
Enge, Ruane & Langlais — "Coverage of a Radio-Beacon Based Differential GPS Network"; *Navigation*, Vol. 34, No. 4.
Kalafus, R. M. "Special Committee 104 Recommendations for Differential GPS Service"; *Navigation*, Vol. 33, No. 1, Spring 1986.
Magnavox — "Magnavox June 1986 Briefing Notes"
Parkinson *et al.* — "Optimal Locations of Pseudolites"; *Navigation* Vol. 33, No. 4, pp. 259–283.
RTCM — "RTCM Recommended Standards for Differential Navstar GPS Service"; RTCM Special Commitee 104, P.O. Box 1908, Washington 20036, March 1988.
Shipmate — "New Perspectives in Satellite Navigation"; Shipmate Holland, 1988.
Stansell, T. A. — "Civil GPS from a Future Perspective"; Proceedings of IEEE Vol. 71, No. 10, 1983. pp. 1187–1195.
Stansell, T. A. — "RTCM SC104 Recommendations for Pseudolite Signal Specification"; *Navigation* Vol. 33, Spring 1986.

Bibliography

GPS and Marine Navigation

Ackroyd, N. A. R. — "Operational Implications of a GPS Navigation System in the North Sea"; Seminar on the Global Positioning System, Univ. Nottingham, 12–14 April 1989

Blood, E. B. — "Navigation Aid Requirements for the U.S. Coastal/Confluence Region"; *Navigation*, Vol. 20, No. 1.

Beattie, J. H. — "Navigation Standards for Merchant Ships"; *The Journal of Navigation*, Vol. 38, No. 3, September 1985.

Beattie, J. H. — "The European Radio-Navigation Mix in the Year 2000"; Proceedings R.I.N. Conference NAV88, October 1988.

Boer, H. B. — "Satellite Navigation for the Merchant Marine"; *The Journal of Navigation*, Vol. 42, No. 5, September 1989.

Bridge, R. W. — "RadioNav 2000 – A European Dimension"; *The Journal of Navigation*, Vol. 41, No. 3, September 1988.

Cockroft, A. N. — "The Circumstances of Sea Collisions"; *International Hydrographic Review*, January 1983.

Cockroft, A. N. — "Development of Routing in Coastal Waters"; *The Journal of Navigation* Vol. 38, No. 1, January 1985.

D'Oliveira, B. — "Rising Stars"; *Yachting World*, January 1986. pp. 122–127.

Hedge, A. R. - "Navigation and Positioning Requirements for Marine Survey Operations"; Proceedings R.I.N. Conference on Global Civil Satellite Systems, 1984.

Igguilden, D. — "The Development of the Latest Survey Technology for the Humber Ports"; The Dock and Harbour Authority, 1986.

Johanneson, R. — "International Future Navigation"; *Navigation*, Vol. 34, No. 9.

Keeler, N.H. - "Future Marine Navigation"; *Navigation*, Vol. 34, No. 4.

Lachapelle, G. — "GPS Marine Positioning Accuracy and Reliability"; *The Canadian Surveyor*, Vol. 40, 1987.

Litgart & Webster — "Position Fixing Requirements for the Merchant Ship"; *The Journal of Navigation*, Vol. 39, No. 1, Jan. 1985.

Machonachie, B. — "Shipping's Space Age Progress"; Safety at Sea, August 1987.

Maybourn, R. — "GPS for Marine Navigation"; *International Hydrographic Review*, July 1984.

Mertikas, Wells & Leenhouts — "Navigational Accuracies"; *Navigation*, Vol. 32, No. 1.

Moskvin & Sorochinsky — "Integrated Navigation and Electronic Chart Display Systems"; *The Journal of Navigation*, Vol. 41, No. 2, May 1988.

Schuffel, Boer & Breda — "The Ship's Wheelhouse of the Nineties"; *The Journal of Navigation*, Vol. 42, No. 1, Jan. 1989.

Sommers & Keubler — "A Critical Review of the Fix Accuracy and Reliability of Electronic Marine Navigation Systems"; *Navigation*, Vol. 29, No. 2.

United Nations — "The Law of the Sea", United Nations Convention on the Law of the Sea.

Van Riet, Kaspers & Buis — "Safety Standards for a 22 metre Deep-Draught Route"; *The Journal of Navigation*, Vol. 38, No. 1, Jan. 1985.

Wells, D. et al. — "Five Years Experience with GPS Navigation"; Proceedings HYDRO 1986 UK; 16–18 Dec. 1986.

Wentzel, H. F. — "Precise Position Fixing for Dredging Operations"; The Dock and Harbour Authority, July/August 1987.
Wingate, M. — "The Future of Conventional Aids to Navigation"; *The Journal of Navigation*, Vol. 39, No. 2, May 1986.

Satellite Communications

Bell, J. C. — "SatComms for All"; *Ocean Voice*, April 1986.
Bell, J. C. — "Satellite Communications and the New Maritime Safety System"; *Safety at Sea*, May 1989.
Bell, J. C. — "Inmarsat Standard C and Positioning"; *Navigation*, Vol. 34, No. 2.
Bell, J. C. — "Mobile Satellite Communications Brought Down to Earth"; INMARSAT Special Publication.
Evans, B. G. (ed.) — "Satellite Communication Systems"; Peter Peregrinus Ltd., IEE Telecommnications series 18, 1987.
Pritchard & Sciulli — "Satellite Communications : System Engineering"; Prentice-Hall Inc., New Jersey, 1986.
Lundberg, O. — "Mobile Satellite Communications and Navigation"; *Journal of Navigation*, Vol. 40, No. 2, May 1987.

CHAPTER 1

Satellites and navigation

1. Global positioning

Introduction

This book will detail the new navigation utility of the Global Positioning System (GPS), in terms of both its applications and its implications to the Professional Mariner and the Shipping Manager. Two Global Positioning Systems are currently under development in the United States and in the USSR. The navigation community can expect to have full uninterrupted access to at least one of these systems by 1993, but with two-dimensional 24-hour coverage in late 1990.

The 1990s will also herald the widespread introduction of a whole host of new technologies to the maritime industry. Many will be based around the familiar themes of integrated ship management and the electronic bridge, some will be newer, such as global position reporting services and electronic fleet management; some will be more radical such as the unmanned bridge and robot ship convoys. All these developments will attempt to cut running costs but still retain the essential requirement of safety during passage. It will be the significant cost reductions realised in implementing the new technology and its widespread application that will characterise the next decade in navigation.

Reliable and highly accurate Global Positioning is certainly not the instigator of all of these changes, but can perhaps be perceived as the glue that holds many of these innovations together. In fact, to come to fruition the integrated ship of the future requires such a catalyst.

In its own right GPS is a highly accurate global navigation service available at minimal cost to the user. Many attributes of the system are going to significantly ease the burden of safe navigation and open up great potential for cost minimisation and increased efficiency. Yet it is mainly in association with the other new technologies of low cost global communications and the low cost personal computer that a 'revolution in navigation' is truly realised.

In the light of this we will not attempt to study the Global Positioning Systems in isolation but will embark on a review of the future of marine navigation in more general terms, using the GPS technologies as a focus. Information will also be presented on the role of satellite communications (SatComms) in navigation, with special reference to the Standard C service soon to be introduced. Although different SatComms systems are to be implemented in the next few years, Standard C, although not fully global, still has many similar attributes to GPS in terms of coverage and scheduling, making the two natural bed-fellows.

2 Satellites and navigation

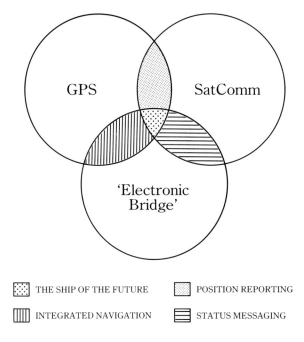

Fig. 1. **The global ship**

In essence this book will deal, therefore, with the global implications of GPS and associated navigation and communications systems capable of operating within this geography. Specific subjects covered will include the applications of the GPS technology to the Shipping Manager; the relevance of higher accuracy GPS to the port and harbour authorities, and the implications of GPS on the safety of navigation especially in coastal waters. In addition, we hope that the knowledge gained through the authors' experience of GPS in a commercial environment will help to target more practical concerns. The sections on the GPS receiver hardware and purchase options will, hopefully, bring the system down to earth.

1.1 The global positioning concept

Global positioning is certainly a new concept and in this light it is hard to consider the American and Russian systems as more than just an evolution in navigation. The earliest forms of navigation, by sun and star, were certainly global, if not all-weather. Currently the all-weather, twenty-four hour navigation potential of the new global systems is under state and military control, which to a certain degree does colour their application. But, as will be detailed in a later section, there are initiatives being taken in satellite navigation by commercial organisations, although these are generally not conceived as being global in nature.

The general term GPS, **G**lobal **P**ositioning **S**ystem, will be used generically to imply both the American and Russian satellite systems. The American system is

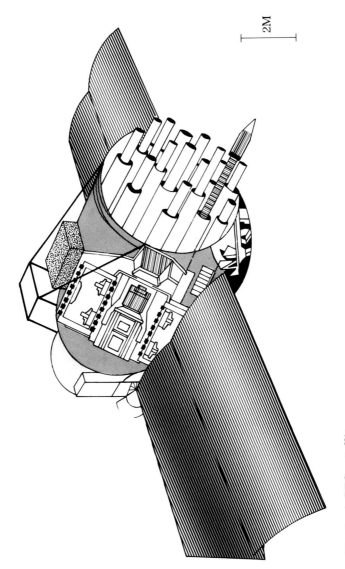

Fig. 2. A GPS satellite

4 *Satellites and navigation*

more precisely called Navstar GPS, standing for **Na**vigation **S**atellite **T**iming **A**nd **R**anging **G**lobal **P**ositioning **S**ystem. The Russian system name Glonass is obtained from **Glo**bal **Na**vigation Satellite System. In actuality there are four electronic navigation systems that could be defined as being global or as of leading to global coverage. These are Transit (Satnav), Tsicada, Navstar and Glonass. All are space-based systems, utilising orbiting satellites as transmitting sources. A fifth, Omega, a ground-based electro-magnetic position fixing system (EMPF), operates at very low frequencies, but does provide near global coverage, if with a reduced accuracy and availability.

What makes the new satellite systems, Navstar and Glonass, unique is their features of high accuracy and high availability, providing position information every second, twenty-four hours a day, three hundred and sixty five days a year. In the simplest mode of operation, accuracies of better than 100 metres (95% level) will be achieved. At the most sophisticated, accuracies of under a metre will be possible whilst under-way. The overriding difference between the accuracies will only be the implementation costs for the user.

The design of both Navstar and Glonass, in terms of the satellite orbits, is specifically to provide a truly global service. Early on in the systems implementation it was thought that the satellites were to be geostationary, i.e. remain in a fixed position in space with respect to the earth's surface. Such an arrangement would have ended up with no coverage in the high polar latitudes, an area of increasing interest to both the military and civilian communities. Recent initiatives undertaken in the Navstar GPS design are intended to further enhance the global performance by increasing the number of satellites in the completed constellation and possibly even slightly change the orbit configuration.

1.2 The navigator's choice

When studying the existing navigation scene, the presence of so many alternative regional navigation systems begs the question as to whether the use of GPS will be as widespread as is being suggested. In assessing the coverage of these existing systems, it appears that most areas where precise navigation is actually required are already well serviced and to similar accuracies as the standard GPS service.

In reality the choice of navigation has often been dictated to the navigator by coverage, availability and accuracy. With GPS there is truly a new choice to be made, but it must offer more than is currently available in order to justify the extra cost. The advantages of GPS will be detailed at length through-out this book, but fundamentally whether or not it is accepted will still be a matter of choice. If anything it is the flexibility of GPS to be anything to any man that sets it apart from the existing systems.

1.2.1 Regional navigation systems

At this stage it is certainly valuable to undertake a short review of existing navigation services available to the mariner, if only to put the GPS technologies into context. At later stages in the book more specific comparisons will be drawn between the familiar navigation systems and GPS.

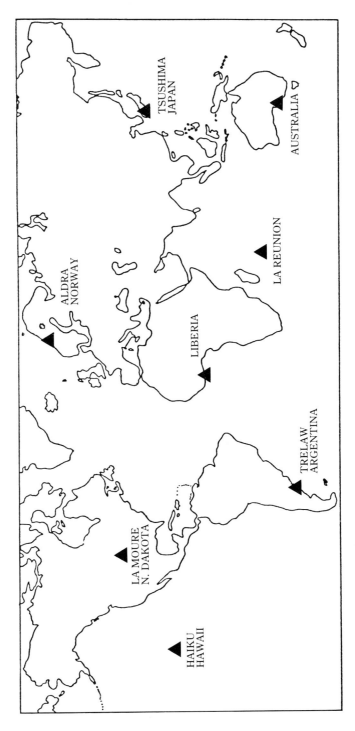

Fig. 3. Omega station locations

6 *Satellites and navigation*

For the next ten years the most realistic picture of future navigation will include the closer combination of both terrestrial and extra-terrestrial positioning systems, especially if the reliance on precise navigation for integrated ship management functions is to be developed fully.

Omega

It is arguable whether Omega should actually be considered as a land-based global positioning system with its ultra-low frequency operation providing positioning information over much of the earth's surface. Omega is based around the use of eight time-synchronised transmitting stations. Observations to these give rise to hyperbolic position lines through phase-differencing techniques, similar to the Decca system. There is a Russian system similar in design to Omega, but little information is available on it.

Omega transmits on three frequencies located at 10.2, 11.33 and 13.67 kHz and their propagation characteristics allow reception of the signals at many thousands of kilometres and even underwater. Omega, similarly to Transit has a significance to submarine navigation, but is also used widely for deep ocean navigation and aircraft. Again no conflict is apparent between the military and civil requirements.

The system provides accuracies at the level of 2–3 nautical miles (95% probability level), although there are proposals to improve these accuracies by using the same differential technique adopted by GPS. The fundamental accuracy of Omega will always be poor, however, due to the propagation characteristics of the low frequencies adopted and the weak chain-geometry over much of the earth's surface. If GPS operation becomes widespread it is likely that Omega may be shut-down soon after the year 2000.

Loran C and Chayka

Loran C and its Russian equivalent Chayka, are medium to long range, low frequency time-difference measurement systems. A master and usually up to four secondary transmitting stations put out a set of radio pulses centred on 100 kHz, in a precisely timed sequence. The receiver measures the difference in arrival time between these transmissions from different stations, thereby producing a hyperbolic line of position based on time difference.

Loran C coverage is fairly widepread, providing potentially very high accuracies, but poor chain geometry, signal propagation effects and restrictions in timing control often degrade its performance. Loran C transmissions can be worked out to ranges over 1500 kilometres providing position accuracies of between 100 and 500 metres dependant on geometry and range (95% level).

A recent accord signed between the USA and USSR has been adopted regarding Loran C and Chayka operation in the Bering Straits. This is to help reduce mutual interference and to embark on a joint test and trials program leading to inter-chain operations.

Loran C is also, at the moment, under consideration for wider-scale utilisation, even though, in the early 1990s outside the USA, its control is to be passed from the

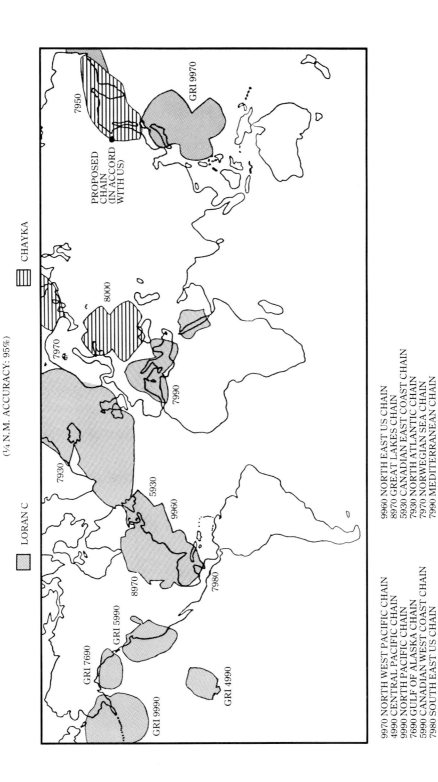

Fig. 4. Loran C and Chayka coverage

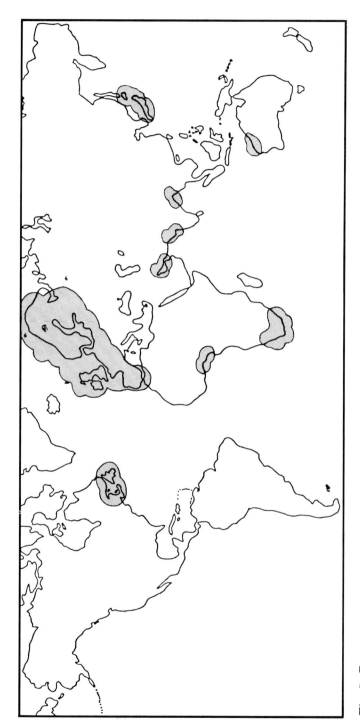

Fig. 5. Decca coverage

US Coast Guard to local national interests. There is certainly a strong lobby to increase the number of Loran chains and their coverage to further complement GPS and provide a necessary terrestrial back-up system. Chain extensions have been suggested for the North Sea and the Bay of Biscay, for example. Whether this will be achieved and at what cost to other terrestrial systems still remains to be seen.

Another initiative being undertaken in Loran C circles is to study closer ways of actually relating the system to GPS, such as the complex issue of using GPS time to control the Loran station clocks. In addition, time synchronisation of adjacent chains would allow for inter-chain fixing, helping to improve overall chain-geometry and coverage. These type of studies are all to try and provide redundancy to GPS operation and to supplement its performance, through the integration of an independent system.

Decca

The Decca Navigator System is one of the oldest of the electro-magnetic radio position-fixing systems, first used successfully in the mid–1940s. It is also undoubtedly the most widely adopted of the position-fixing systems in Western Europe at the current time. Forty-four Decca chains are in operation, concentrated around the European Continental Shelf, but the system is also present on four out of the five continents.

Decca works through taking observations to pairs of transmitting stations again using phase-differencing techniques to give rise to hyperbolic lines of position. At least four stations, usually known as the Master, Red, Green and Purple, transmit a continuous wave on different frequencies in the 70–130 kHz frequency range. The receiver measures the difference in phase angle between the transmissions of the Master and one slave station. This difference in phase is effectively a distance-difference measurement, although with some ambiguity, which needs to be solved by special lane identification transmissions.

Decca operates at a shorter range than Loran C, with reliable 24 hour positioning up to ranges of about 240 nautical miles, giving accuracies of between 50 and 200 metres (95%) in good-to-fair geometry. At longer ranges over 240 nautical miles, the night-time performance is degraded severely, with day-time performances achieving figures in the order of 400 metres.

1.2.2 The satellite alternatives

As well as the traditional terrestrial navigation chains there is now also a new generation of regional satellite-positioning systems. These are not designed to have, nor are capable of, complete global coverage, but they do offer increased competition to the regional navigation services and at undegraded accuracies in the order of five to twenty metre (95% confidence). Coverage by geo-stationary orbiting satellites will always degrade severely above and below 70 degrees North and South, also exhibiting weakened geometry around the equator itself.

These systems are given the general name of Radio Determination Satellite Systems (RDSS) and are generally conceived as ancillary navigation payloads, loaded onboard communication satellites. RDSS systems are primarily geared to the potentially massive market of land navigation, from emergency services to truck operators, security firms, taxi services and even the private car. For these purposes some contain an inherent data transfer service and position reporting element. They are all operated in the commercial sector.

Star-Fix

Star-Fix is a regional satellite positioning service operated in the Gulf of Mexico and Mid-Western United States by PANAV Inc, a Joint Venture between John E. Chance and Associates and Analytical Technology Laboratories. It utilises a technology basically similar to that of the GPS systems, with a Master computing facility in Texas controlling ten remote tracking stations monitoring the satellite's position, which are in geo-stationary orbit. Position data for the satellites is uplinked at 6 gHz and the carrier downlink is at 4 gHz. The Star-Fix satellites are primarily communications satellites located in the equatorial plane.

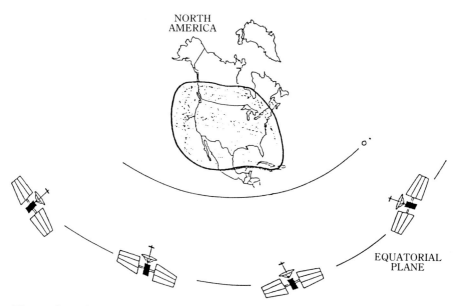

Fig. 6. Star-Fix system coverage

Star-Fix utilises spread spectrum techniques like GPS, with pseudo-random codes being used to measure radio travel time from the satellite to the user. Again it is a passive system. The satellite position data is transmitted down superimposed on the carrier frequency. Due to the geometry of the satellites Star-Fix can only position in two dimensions, but it can achieve this to an accuracy of better than five metres. Star-Fix has already been successfully incorporated with a differential Navstar GPS service to provide three dimensional positioning and system redundancy.

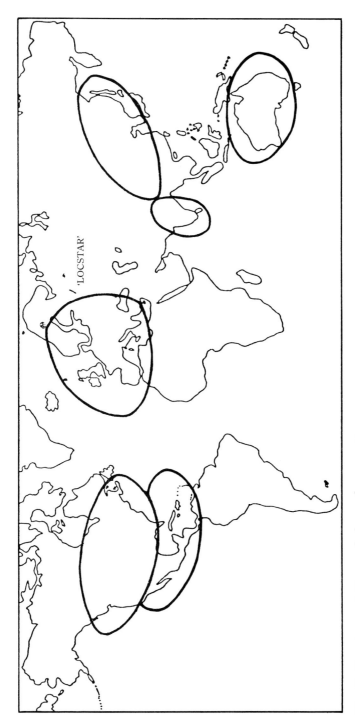

Fig. 7. Geostar: existing and planned coverage

Due to the low gain of the received satellite transmissions Star-Fix operates with a remotely steered four-element tracking antenna. This is fairly large in size, equivalent to an INMARSAT Standard A antenna, but recent developments are helping to reduce this size.

Star-Fix is currently operated in a full service environment with the main customers being connected with the oil industry in the Gulf of Mexico. Day rates are currently high and unlikely to appeal to general navigation users for this reason. In this respect they equate with the higher accuracy survey packages already operated for the exploration community. Yet, as user days increase and the economies of scale come into play, substantial reductions can be expected. Even so, whether such advertently commercial services will ever become general navigation tools remains to be seen.

Geostar

Geostar, operated by the Geostar Corporation, is primarily a position-reporting service that currently integrates existing terrestrial Loran C positioning with satellite communications technology. However, it is also designed to have an integral satellite ranging capability, it is hoped that this service will be introduced in the early 1990s with a similar timescale to GPS. A European equivalent, known as Locstar, is planned for introduction in the mid 1990s.

Much of the technology behind the Geostar system has been developed around the data transfer and messaging capabilities of the system. Its independent position-determination capability is still under development, but it does differ substantially from GPS and Star-Fix in that it is an active system. In the Geostar concept the position calculation is actually undertaken effectively by triangulation and at the Central installation, not in the transceiver.

The position calculation process begins with the control centre sending out coded timing marks many times per second. Upon receipt of these the user responds with a set of very short coded transmissions of between 20–80 milliseconds, also containing data. This transmission is routed independently through at least two satellites back to the central computer. This gives effectively at least two different radio travel times. This information is used to determine position at the central processing centre which is then routed back to the user's transceiver. Like Star-Fix, Geostar can only give two dimensional position information. Due to the processing power of the central computer Geostar claim there is practically little limitation to the number of users of the system.

Proposed systems

A number of proposed RDSS systems are currently under investigation with some actually reaching the stages of initial financing. Although it is obviously difficult to foresee how many of these will take off (literally!), their proliferation is indicative of the future importance of satellite navigation. Most of these systems propose the use of communications satellites such as the new INMARSAT space segment and certainly developments along these lines can be expected to materialise.

NAVSAT

One of the most well researched proposals that could simply be expanded to a global service, is a planned European satellite navigation system referred to as NAVSAT. NAVSAT studies were undertaken by the European Space Agency alongside another satellite positioning initiative known as GRANAS, conceived in West Germany. The two analyses were married together in early 1987 to form a combined single European proposal. NAVSAT takes the unique concept of unifying geostationary satellites, such as the communication satellites, with high earth-orbiting satellites such as those used in the Navstar GPS program. This has been called GEO/HIO mix [Rosetti, C., Diederich, P. *Trends in The Evolution Of Global Satellite Navigation Systems*].

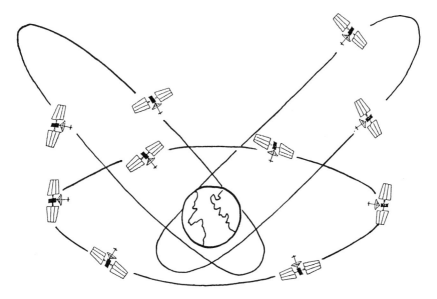

Fig. 8. Three out of seven planes in a GEO/HIO mix

The development concept behind NAVSAT is to instigate it on a regional basis, satisfying areas of high demand such as Europe, until, if necessary, global coverage is achieved. This can only be attained by using the high earth-orbiting satellites similar to the GPS satellites, which alongside elliptical orbits will provide a more regional coverage. The advantage behind incorporating geostationary satellites is the major cost savings in the expensive space-segment element to such systems, as with the Star-Fix and Geostar systems, the navigation payload will hitch a ride with the primary communication package.

The provision of such a system as NAVSAT would give the civilian community the ideal tool for general navigation. The control of the system would be firmly in civilian hands and could be developed alongside multi-national, multi-interest requirements. If this system ever comes to fruition then, alongside Navstar GPS and Glonass, the future of global navigation will be determined.

14 *Satellites and navigation*

2. The global positioning systems

2.1 The first satellite systems

Transit

Satellite Navigation was first conceived after the launch of Sputnik 1 in 1957 when scientists realised that the small bleeps emanating from this first space vehicle could be used to locate a point on the earth's surface. But the first truly global satellite navigation system did not come into being until the early 1960s with the installation of the Navy Navigation Satellite System also known as Transit (SatNav). This came from the same military origins as the Navstar system, so a lot can be learned from tracing its implementation.

Fig. 9. Transit orbit arrangement

Transit works on the Doppler principle, using six low orbiting satellites and two transmitted frequencies of 150 MHz and 400 MHz. In brief, position is achieved by measuring the change in frequency of the satellites transmissions as it speeds past in low orbit. Having information regarding the satellites position and velocity allows the computation of position to be made through counting the range rate or accumulated cycles of the Doppler frequencies. More cycles are received, i.e. the frequency is increased, when the satellite and receiver come closer together.

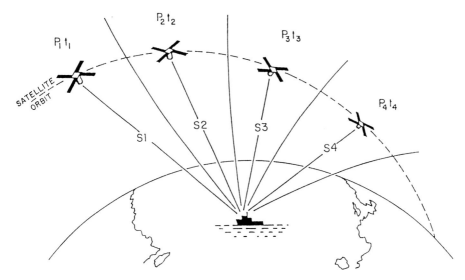

Fig. 10. Transit system principle

With the receiver counting this range rate and with knowledge of the source satellite frequencies and the subsequently, shifted, received frequencies, it is possible to calculate range. Applied to the position of the satellite this gives one line of position. An intersection of a number of lines of position, to define a point, can only be achieved by making a sequential number of observations of the satellite through its orbit pass, in effect building up a history of position lines over a period of time. This is why, for the highest accuracy it is necessary to interface a speed log and a gyro to a Transit, to keep track of the vessel's movement through this window.

A Transit fix, taken underway, can only be made in two dimensions (latitude and longitude) and has an accuracy of not much better than 250 metres (95% level). But the overiding limitation of the system is in its coverage. On average, a fix can only be made about every 1.5 hours, the interval between satellites being in view. Even so, for general navigation purposes out of sight of land and shore based EMPF systems, this is still of significant value, especially when used alongside dead reckoning.

The Transit principle has been so successful over the last twenty-five years that a new updated version is actually being considered in commercial circles. This would utilise a host of very low cost, small, satellites put up in low orbit. The main difference is that there would be many such vehicles launched to provide almost continuous coverage.

2.2 Satnav and GPS: a perspective

In exactly the same way as the Glonass and Navstar technologies mirror each other, there is also an equivalent Russian doppler satellite system known as Tsicada. This has been in use since the late 1960s and performs an identical function to Transit, both primarily introduced for the navigation of submarine fleets. A Transit fix would be used to update inertial navigation systems used onboard the submarine for primary navigation. In essence all global positioning systems up to date have been designed primarily for military purposes.

Navstar GPS and Glonass come from the same stable as Transit and Tsicada and if this analogy is continued there would appear to be little to fear regarding the operational reliability of the new systems. Although ultimately the on/off switch is in the hands of the military community, this has rarely caused a problem in the history of SatNav. During the last five years of partial GPS coverage there have undoubtedly been some periods of disturbance and down-time, but this was exactly the same with the early testing period of the Transit system.

The American Department of Defense are very open about the fact that the system is currently in a development and testing phase and are so concerned about current reliance on the system that they actually pre-publish any intended down-time or testing windows. In 1995 the Transit system is scheduled to be switched off. By this time, in the authors' opinion, there is little doubt that Navstar GPS will be a fully-completed, proven and reliable navigation service.

The one military concern has been the provision of the highest level of accuracy to all users in stand-alone mode. This accuracy could potentially be a military threat if available to unfriendly users. The 10–15 metre capability is, therefore, not to be provided in stand-alone mode, limited by the implementation of a deliberate degradation known as Selective Availability (S.A). However this degradation, resulting in the 100 metre horizontal (95% level), can be removed relatively easily by a differential technique. This uses a reference receiver sited at a known position to define the errors, the transmission and near real-time application of these figures to a mobile user, then effectively removes the effects.

This technique does not seem to concern the military authorities who actually appear quite happy about its use. Possibly, as it relieves the pressure from them for the release of the full accuracy potential of the system. Again what is interesting in all this debate is the actual concern of the systems' operating authorities to provide a usable service to the civilian community.

Certainly, pressure from the American Congress would not allow the system to be utilised for purely military benefit, especially considering its 12 billion dollar price tag. Like Transit, Navstar GPS has a civil role and has been designed and authorised with this utility in mind. The frequent posturing by certain sections of the navigation community regarding its military origins and accountability is often for less than altruistic reasons.

The global positioning systems 17

2.3 NAVSTAR GPS—a system description

The technology behind both the Navstar GPS system and the Glonass system is necessarily quite complicated and in respect of this a much fuller explanation of both systems is given in Chapter 5 of this book "The GPS Detail". In this section we will limit the explanation on the Navstar system to a more introductory level providing the necessary information to the more casual reader interested in the applications.

2.3.1 The system design

The Navstar GPS system design consists of three integral parts; the Space segment, the Ground/Control segment and the User segment. All these parts operate in unison, providing high accuracy three dimensional positioning data, velocity and accurate time to suitably equipped users, world-wide.

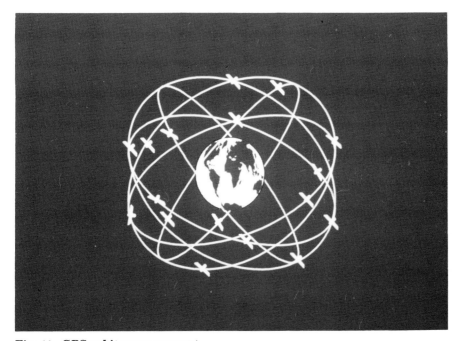

Fig. 11. GPS orbit arrangement

When completed, the system will consist of twenty one satellites with three active spares, although officially an eighteen satellite constellation with three spares is the design profile. The increase to a 21 + 3 constellation is to occur as soon as practicable. The satellites are to be placed in very high polar orbits, just over 20,000 kilometres up and have an orbit period of 12 hours. The final constellation will provide complete four satellite coverage world-wide at all times within the next decade. Considering the present launch schedule this is not to be expected before 1993/1994. Full two dimensional coverage, suitable for most marine applications, can be expected by 1991 and possibly earlier if existing development satellites stay operational.

2.3.2 Current configuration

At the time of writing six Block 1 (prototype) satellites and five Block 2 (production) satellites are in orbit. Of the original Block 1 satellites two are close to failing and are unlikely to provide many more months of service.

At present, coverage is limited to about eight to twelve hours per day dependent on location. This will increase markedly over the next years as more satellites are launched, filling in the holes. Even with the completed constellation there will still be periods of degraded performance in some parts of the world. This will be due to short spans of weak geometry. Figure 12 illustrates the partial coverage experienced at the moment in the English Channel.

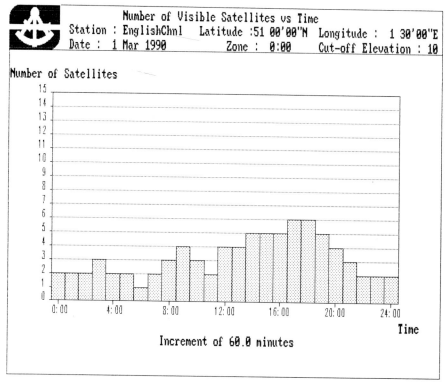

Fig. 12. English Channel GPS Coverage as at March 1990. This is subject to continuing addition. (Courtesy Trimble Navigation Ltd.)

2.3.3 Navigating with GPS

Each satellite, for navigation purposes, transmits a unique coded sequence allowing identification of the satellite, the calculation of ranges to it and the ability to decode data from it. The Navstar satellites transmit on two L-band frequencies centred on 1575.42 MHz and 1227.60 MHz. These are referred to as the L1 and L2 frequencies respectively and are based on a fundamental clock frequency of 10.23 MHz. Each signal has a sequence superimposed on the carrier frequency by modulation techniques.

These modulations are in the forms of codes, a Precision (P) code and a Clear/Acquisition (C/A) code. These are alternatively called the Precise Positioning Service (PPS) and the Standard Positioning Service (SPS). The L1 carrier has at present both code modulations, whereas the L2 contains only the P code. This is not expected to change in the future. It is also understood that at the time of the completed constellation the P code will be witheld from nearly all civilian users and further encrypted to the Y code. The P code is more precise than the C/A code with a smaller code bit interval, yet the performance difference between the two was not as significant as the system designers had expected. This is one of the main reasons for the introduction of Selective Availability.

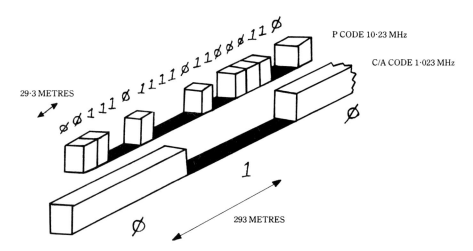

Fig. 13. P Code and C/A code modulations

The code modulations help to provide resistance to interference or deliberate jamming by spreading the signal out over a wider bandwidth. They also allow relatively low power transmissions to be used, as each satellite code has a unique sequence easily identified by statistical means and is difficult to confuse with background noise or the signals from each other. The code is modulated on the carrier by changing its phase, thereby breaking up its sine wave form. The use of codes is another alternative to the use of continuous waves in Decca and pulses of energy in Loran. All effectively allow distance difference measurements to be taken.

The rationale behind transmitting two frequencies is that, by being multiples of a fundamental frequency (10.23 MHz), a specific relationship can be expected between the two. Any disturbance to this relationship, such as refraction delays introduced by the earth's ionosphere, may therefore be determined by studying the different effect it has on the two separate transmissions. Two frequency operation becomes most critical during periods of high sunspot activity such as that currently being experienced and expected to last until 1992.

Pseudo-range measurements

There are two fundamental observations that are made to the GPS satellites that can be used to determine position. These are the pseudo-range and the phase observable. Pseudo-ranges allow position to be calculated under all dynamic and static conditions, whereas phase observations have some limitations to their use.

A pseudo-range is essentially the radio travel time between the satellite and the receiver, expressed in metres. This is obtained by decoding the P or C/A code which, in essence, contains a snapshot of the satellite clock at the time of transmission. This is compared to a receiver clock at the time of reception, thereby giving a time/distance measurement.

In actuality the measurement procedure is more complex. The pseudo-range is determined through the receiver, generating a replica code to that transmitted by the satellite. This replica code is matched to the incoming satellite code and the amount of delay that the receiver must apply to its code indicates how much they are out of step. This mis-match in time is a function of two separate sources of time differences.

The first difference is due to the lack of synchronisation between the receiver and satellite clocks. Essentially they start off by telling different times. This is the same on all the measured ranges and can therefore be mathematically calculated. The second time difference is because the satellite and the receiver are in different places and so it takes time for the transmission to move between them. This second time difference is obviously the range.

The first time difference, more correctly termed the clock bias, can be calculated but it does initially present another unknown for the receiver to solve in addition to position. The presence of this clock bias explains why the measurement to the satellite is a *pseudo-range* as opposed to a range.

Moving on from this, for the position of the receiver to be solved it must also know the position of the satellites. This information is also transmitted down from the satellites as a formatted navigation message in addition to the codes. This contains all the orbital information necessary to calculate the satellites position at the time of the range measurement and any clock corrections necessary for the satellites themselves. Further information regarding the performance of the satellite and data for modelling ionospheric delays is also included. All together this information is known as the satellite ephemeris.

Accurate GPS measurements are therefore dependent on the precise transfer of time between the Ground/Control segment, the Space segment and, ultimately, the User segment. It is the reponsibility of the Ground/Control segment to maintain a common time base across all the satellites and to provide accurate data as to their position in space at all times.

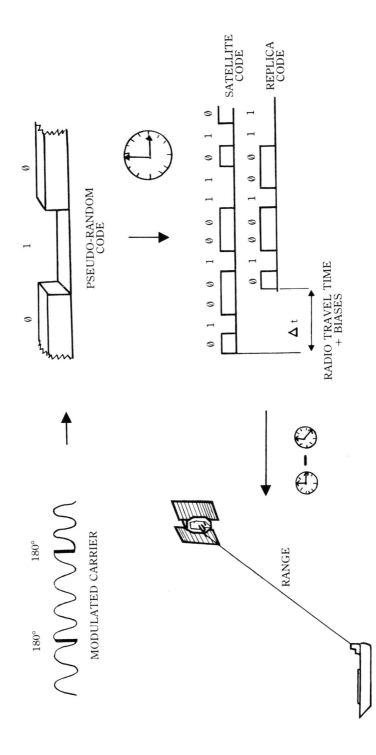

Fig. 14. The pseudo-range

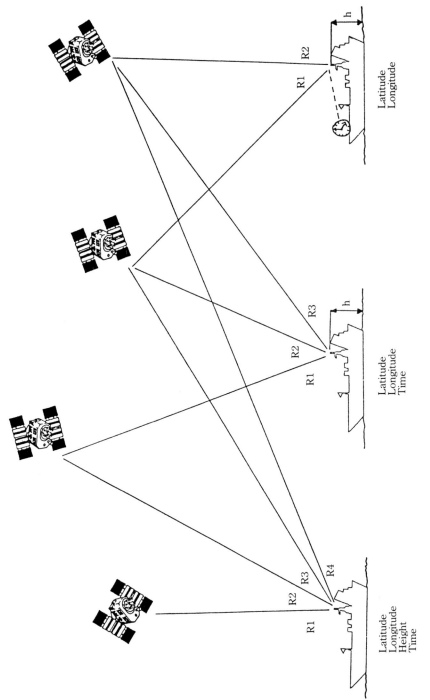

Fig. 15. GPS and position aiding

The calculation of position

As in terrestrial radio-positioning systems, more than one range or hyperbolic line of position is required to produce a unique point on the earth's surface. For calculation of position using GPS four unknowns require solving, X,Y,Z and T the three dimensional space co-ordinates (position) and the clock bias. To calculate this unaided would require four satellites to be observed, giving four ranges to solve for the four unknowns. This number can be reduced by solving some of the unknowns for the receiver prior to the calculation.

If height is known accurately, as would be the case on a ship, then only three unknowns remain, requiring observations of only three satellites. Likewise if the receiver clock bias could be determined independently then only the three position unknowns remain. If both height and clock aiding are used then only two position unknowns remain and only two satellite ranges are required to solve them.

These approaches can be used to give redundancy in range measurements (more ranges than are actually needed) resulting in a higher confidence, but are of more value at the moment to extend the workable satellite coverage. For height aiding, height should be known accurately to within a few metres and clock aiding requires the use of an expensive rubidium atomic frequency standard. Height aiding is especially valuable as it helps to improve geometry as well.

2.3.4 GPS system accuracies

Great care is needed when assessing the accuracy of any system as it is a function of many different sources—the user to station geometry, the noise characteristics and resolution of the frequency, range separation, system monitoring tolerances and many more. With GPS there is a new consideration which is the deliberate degradation of the system for military needs. This is known as Selective Availability or more succinctly as Accuracy Denial. There is little to distinguish between the results obtained on the two codes without Selective Availability (S.A), one of the main reasons for its introduction.

All commercial receivers available on the market are designed to select a configuration of satellites with the best geometry and as such provide optimum positioning within the constraints imposed on them. As mentioned geometry can be improved by increasing the number of satellites used or reducing the number of unknowns.

The satellite geometry is presented to the user by a factor known as the Dilution of Precision. These figures are used to assess the potential positioning quality of a certain satellite constellation and to help provide realistic quality control information. The procedure used to define these values is quite complicated, but it relates the difference in three dimensions of the user to all the considered satellites in a geometrical sense. (See Fig. 16 on p. 24.)

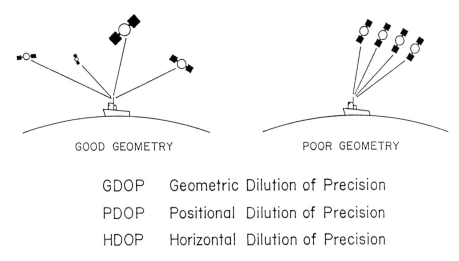

Fig. 16. Geometry effects

The resultant DOP figure then suggests the amplification of pseudo-range measurement error into user positioning error. Different DOP's are used dependent on the type of position being calculated. Horizontal Dilution of Precision (HDOP) is used for a two dimensional fix, Position Dilution of Precision (PDOP) for a three dimensional fix.

DOP figures are actually used by a navigator in the following way. If a composite pseudo-range measurement error of 10 metres (95% probability) is assumed for the system, then multiplying this figure by the relevant DOP value, e.g. 3.0, gives an overall positioning accuracy of thirty metres. DOP figures are therefore only relative numbers, with smaller DOP's giving better accuracies. The probability figures detailed in the circular brackets are indications of confidence in the position accuracy.

Table 1. The dilution of precision

GDOP	Geometric dilution of precision Integrates X,Y,Z and time	EDOP	Easting dilution of precision
		NDOP	Northing dilution of precision X and Y positioning separated
PDOP	Position dilution of precision Integrates X,Y,Z, three D positions	TDOP	Time dilution of precision For time transfer users
HDOP	Horizontal dilution of precision Two D marine positioning		

The final constellation is designed to guarantee PDOP figures of better than 12.0 all over the earth's surface, though figures of better than 7.0 are to be generally expected. For maritime applications HDOP figures are more useful and the design figure of 3.0 is to be expected for the majority of coverage.

Single receiver GPS accuracies

All accuracies quoted in the following section are quoted at the 95% probability level. To relate these to other published figures or those indicated in receiver brochures please read Chapter 5, section 4.3.

If the system were left undegraded then trials results over the last five years indicate that an accuracy of thirty metres in two dimensions is readily achievable under average geometry conditions. This does not degrade with range nor is it subject to nighttime or weather disturbances. In fact if anything GPS is more stable during the night.

However, the standard positioning service offered to the civilian sector will be compromised with selective availability. This will provide a one hundred metre accuracy, though probably a little better in the horizontal component if height aiding is used. Even at this level, considering its global availability, it will significantly improve most vessels' navigation capabilities.

Enhanced accuracy levels

Certain techniques can be adopted even within an environment of selective availability to further improve on the real-time system accuracies. Primarily this is through the use of differential techniques. A simple differential service using low cost receivers and low cost VHF radio transmitters can be expected to provide accuracies at the ten metre level, a significant improvement over the hundred metre standard service. Corrections to the satellite as seen at an on-shore monitor are transmitted to a mobile user, thereby improving his accuracy.

There are also even more sophisticated methods available. If observations can be made on the actual carrier frequency of the GPS satellites then the pseudo-range can be made even more stable. This is achieved by taking doppler measurements, similar to the Transit approach, and using the velocity information determined from this to smooth the pseudo-range. In differential mode real-time accuracies of under five metres can be achieved with this technique. Current trials and tests suggest that with even more sophistication (and cost) accuracies of under one metre may well be possible for a moving vessel.

This level of accuracy is probably of little concern to the navigator but may well have applications in related services such as harbour surveys and dredging operations. These topics are covered more fully in Chapter 4 of this book.

26 *Satellites and navigation*

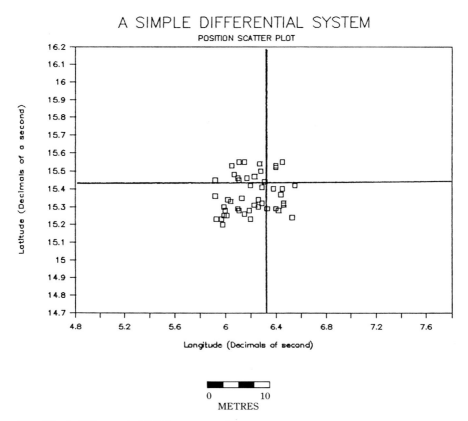

Fig. 17. A differential GPS scatter plot

CHAPTER 2

GPS and the ship

Introduction

This part of the book will deal with the implementation of GPS on the ship itself. It will review the implications of an accurate global positioning system on the bridge and its role in realising the truly integrated ship. On a more practical level it will detail the GPS hardware that a Master and Duty Officer will have to operate, but, more importantly, will at some point need to select. In light of this the section The GPS receiver covers the different types of receivers available and focuses on the critical features of a GPS receiver to a user.

1. GPS and the electronic bridge

1.1 The integrated ship

As the drive for greater efficiency in shipboard operations continues to gain pace, particularly amongst the fleets of the industrialised nations, the level of automation onboard increases. Where once the different ships' departments (Bridge, Engine Room, Radio Officer etc.) were viewed as independent and separate domains, they are now often considered as interdependent parts of the overall ship system.

Until quite recently the introduction of electronic equipment into any of the ships' departments reflected the historical separations. There was at best little effort, and more often no effort, to pass on the benefits of increased efficiency and better levels of information from one department to another. Historically, ships' bridge electronic systems have been introduced to replace or assist with specific, separate functions and duties of the Officer of the Watch and watch personnel. For example the automatic pilot replaces a helmsman, and fitting an Electronic Position Fixing System (EPFS) such as Decca or Loran C assists in position determination. Although such electronic systems usually improve the efficiency and/or performance of a particular function or functions, they have also undoubtedly increased the workload of the duty officer. This becomes critical during certain stages of the ship's passage such as making landfall and navigating coastal waters.

One of the reasons for this increased workload of the Officer of the Watch is the lack of integration between the various systems he has to operate on the bridge. He must first know how to operate a significant number of different systems, for example

EchoSounder, EPFS, Radar, Autopilot, ARPA, Radio and/or Satcom, all of which may very likely change from ship to ship. He must also have a good understanding of the accuracy and potential errors inherent in each system. GPS now becomes another tool the bridge officer must become familiar with alongside the potential applications of differential operation, position reporting services and the like.

The task of watchkeeping on a non-integrated bridge involves physically moving between the various systems, extracting the desired information at the correct time intervals, assimilating information from different sources (including most importantly the chart) and making decisions based on the information at hand with due regard to the prevailing conditions and the international collision regulations. Integrated Bridge Systems are currently being investigated by ship-of-the-future projects in West Germany, Japan, France, and the Netherlands. The objectives are an ergonomically designed bridge suitable for one man operation, and to include engine room control and monitoring. These systems are generally seen as part of an integrated ship system with the final objective, certainly in the case of Japan, of developing ships capable of unmanned operation.

The integrated ship is obviously beyond the scope of this book, however GPS does have an important role to play in the development of integrated bridge systems, and the concept of totally automated ship operation, for example, is only feasible with the existance of an all-weather 24-hour Global Navigation System. The drive for automation is, of course, cost-related with crew costs one of the

Fig. 18. Artist's impression of the integrated bridge, including 2690 BT ARPA, CVP 3500 and MNS 2000 with LSR 4000 screen (courtesy Racal Marine Electronics Ltd)

largest elements in a ship's operation. The concept of Robot Convoys, again being pioneered in Japan, is being developed on the assumption of reliable global positioning and communications. Even within an environment of Selective Availability the relative position of the master and slave ships can be accurately known to ten metres or so, through the application of relative differential techniques.

Some integration of bridge functions has already taken place with the introduction of Integrated Navigation Systems (INS) and Engine Monitoring and Control Systems (EMCS). Such systems group together certain bridge functions to improve the efficiency and ease the work-load of the officer of the watch. However, they are a result of evolutionary pressures in bridge design and have been developed with what could be described as a "bottom-up" approach. This contrasts with the "top-down" approach which would have accompanied the development of such systems if they had been viewed as sub-systems of the integrated bridge rather than individual stand-alone products.

Exercises in fuel efficiency have shown that, with careful engine control and monitoring substantial cost savings can be made in an integrated environment. This type of information can be displayed to the Officer of the Watch through a live situation report which provides all the relevant information on the one screen, such as course steered, engine revolutions requested and actual, speed, passage economies etc. Fundamental to the successful application of all this information is the accurate knowledge of position and change in position; distance to waypoint, course to go, time to go and distance through the water. Such systems have been

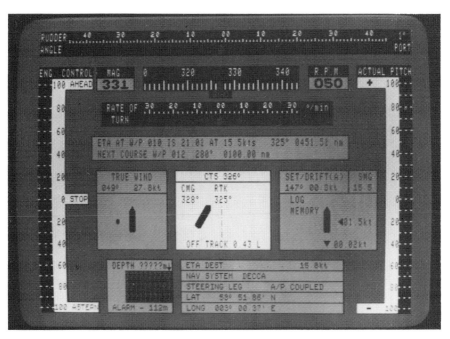

Fig. 19. Typical display of LSR 4000 Live Situation Report (courtesy Racal Marine Electronics Ltd)

30   GPS and the ship

shown to provide savings of up to ten percent on expended fuel. This can be many thousands of dollars on long ocean-going routes.

The bottom line, though, is that all this technology costs money. Possibly hundreds of thousands of pounds are needed to install such a system. It is therefore important to the ship owners that if such as system is to be installed it needs to be fully utilised. In most instances full utilisation will require continuous, accurate positioning throughout the ship's passage. Unless the vessel follows a repetitive route, such as a RoRo ferry, through existing navigation coverage, then a Global Positioning System becomes critical to the practical realisation of such advanced systems.

For example, the Racal Marine Electronics Live Situation Report System 4000 integrates a whole host of electronic navigational and engine monitoring aids. Two interswitched colour rasterscan radars operated as an X-band general navigation and anti-collision unit and an S-band ARPA, a colour video trackplotter and an adaptive autopilot are integrated in a central computer console. In addition an echo sounder, magnetic compass, wind and speed indicator and a speed log are also interfaced. But central to all this equipment is the MNS 2000GL navigation receiver. This allows the optimum selection of navigation system from a choice of GPS, Transit, Decca, Loran C and Omega. The degree of choice indicates the essential nature of positioning to such an integrated system. No doubt as GPS coverage increases it will become the mainstay of such advanced systems. In fact, it is difficult to envisage their widespread adoption without this utility.

Fig. 20. The Racal Decca MNS 2000 on board Sealink Ferry Hengist

If the more accurate GPS differential service is available to a ship then this increases the potential efficiency of such a system to a level where an onboard central computer could be monitoring the ship's performance to the highest levels. This could be expanded to the degree that in effect real-time ships' trials are being carried out, allowing a performance model of the ship to be built up over time under all sea and weather conditions. This can only help optimise ship operation.

1.2 GPS and the operator

One major concern with the implementation of all this technology onboard the vessel is the operators. The type of training now required has appeared to have moved away from the more traditional maritime skills and training towards that of a systems analyst/computer operator. In terms of GPS on its own, the training element is not as complicated. Most commercial GPS receivers are relatively easy to operate and similar in protocol to most existing navigational equipment. It is always a good idea, though, when the equipment is first installed, that all likely users are present to be instructed in its operation. More importantly perhaps, they should ask questions regarding the equipment and the system in general. Suitable questions may be suggested from the following section on the GPS receiver hardware. It is not essential to the operator to understand fully how the system operates, but a working understanding does help, especially to identify when something has gone wrong. If GPS, however, does become the core of developments such as the integrated ship or becomes important in terms of traffic management schemes and maintenance of safe separation, then a greater understanding will be required of this integrated operation.

At present the impetus of training is coming from the commercial sector which is trying to introduce the new technologies, and courses subsidised by equipment purchase are probably the mainstay of training. Universities and polytechnics are also developing courses in these fields with quite advanced ship simulators now available to long stay and short stay students. These institutes are also very aware of the commercial advantages of offering short courses on GPS, for example. Over the next few years these are likely to become even more frequent.

2. The GPS receiver

2.1 The GPS receiver types

Broadly speaking GPS receivers break down into three main types:
1. Parallel/multi-channel receivers
2. Slow or fast sequencing receivers
3. Multiplexing receivers

Parallel/multi-channel receivers

These tend to be the most expensive of the GPS receiver types and are generally used for higher accuracy applications or in high dynamic situations. A marine parallel/multi-channel receiver would currently be priced at around $US 25,000 and should include many sophisticated navigation functions. A parallel receiver has a dedicated channel assigned to each satellite and as such often uses in excess of eight integral channels. Such receivers also generally utilise one of these channels as a rover to speed up initial satellite acquisition and download ephemeris data. This dedicated approach gives access to continuous, uninterrupted measurements. This is extremely important for tracking satellites under high dynamics as in an aircraft.

This type of design is also critical if successful measurements of the carrier phase observable is required. This is necessary for the centimetric accuracies obtained by the land survey techniques and increasingly the one or two metre accuracies obtainable through high accuracy differential methods. Parallel receivers also offer the best signal to noise performance and, as such, the most stable pseudo-range measurements in their own right. Current technology allows the reduction of a channel to a single chip where, initially, a board per channel was the norm. With multiple unit cost reductions it is now possible to have a twenty-four channel unit able to track all visible satellites on both frequencies for the same price as the first dual channel receivers sold in the mid 1980s.

Slow and fast sequencing receivers

Historically these have been the receivers designed for the marine navigation market. They utilise a much simpler hardware architecture, resulting in somewhat reduced performance but a substantially lower cost. A small dual channel fast sequencing receiver should cost somewhere between $US 5,000 and $US 10,000. These receivers contain either single, or, more commonly, dual channel tracking capability. In the dual channel models the second channel is often used as a rover, adopting the house-keeping roles of initial satellite acquisition and ephemeris downloading. The other tracking channel is used to sequence between the available satellites taking measurements, although this is usually kept to a minimum of four tracked satellites. For a fast sequencing receiver this procedure is normally achieved in just a few seconds. Slow sequencers may take in excess of ten seconds. A fast sequencing receiver may also be able to track the carrier, though it will be more susceptible to loss of carrier lock under higher dynamics. For most marine applications two channel fast sequencing receivers are perfectly suitable.

A single channel receiver tries to undertake all tracking and satellite management functions with just one channel. This actually results in a break in the measurement data under some conditions. For example, when it downloads ephemeris. Single channel receivers tend to be slow sequencers and are becoming much less common in the market place, with little to recommend them. The main weakness with sequencing receiver, especially exacerbated in the slow sequencers, is that within the sequencing cycle estimations have to be made to satellites currently not residing in the tracking channel. This introduces additional measurement error also making the process unsuitable to high dynamics where a substantial change in the vehicles attitude between satellite updates might make reacquisition difficult. This slight increase in noise is generally removed by a statistical filtering technique known as Kalman filtering.

Multiplexing receivers

These receivers offer a half-way house between the sequencer and the parallel multi-channel receiver. Although similar in concept to the sequencer, more sophisticated design allows the receiver to switch between all tracked satellites in under twenty milliseconds. This means, for example, that the receiver is capable of taking measurements to five satellites within one bit of the satellite message. This gives the impression of providing continuous, bounded tracking in similar form to a multi-channel receiver without the need for dedicated channels and inter-channel calibration.

Multi-plexing receivers may have a limit to the maximum number of satellites tracked without becoming essentially just fast sequencers. In addition this technique does suffer from having less signal gain for each satellite as they are only tracked for a very short period. This may result in slightly poorer performance under marginal tracking conditions (low signal to noise) than, say, a multi-channel receiver. They do perform generally better in dynamic conditions than sequencers and allow integrated doppler observations to be taken, which help smooth the pseudo-range. They may cost slightly more than a sequencer.

Fig. 21. A dual channel fast scanning receiver (courtesy of Trimble Navigation Ltd)

34   *GPS and the ship*

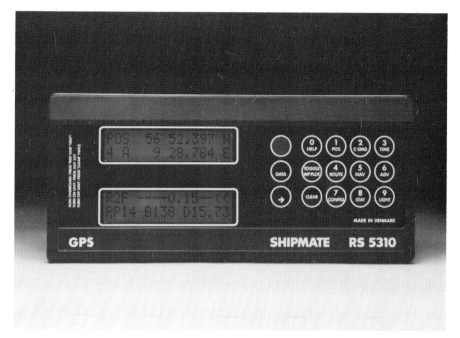

Fig. 22. Shipmate RS 5310—a multiplexing receiver

2.2 The hybrid GPS receiver

An important development in navigation technology has been the integration of the GPS receiver with existing navigation systems equipment. This has given rise to what is commonly called the Hybrid GPS Receiver. This equipment is capable of making observations to both the new satellite system as well as to existing and trusted navaids. The hybrid receiver is certainly a powerful and worthwhile tool for the navigator and should be considered carefully as a suitable GPS purchase option.

The concept was primarily introduced to allow the coverage gaps caused by the partially completed Navstar GPS system to be bridged using existing navigation signals. Not only does this allow GPS receiver manufacturers to sell their equipment well in advance of a completed constellation, but it gives the user access to the new technology alongside a proven and understood tool. This area of built-in confidence must be seen in these formulative days to be of great value. Equipment manufacturers had already appreciated the advantage of such mixed receivers with units such as the Racal MNS2000, already well established in the market place. This receiver already had the capability to track the Transit satellite system, Omega, Loran C and, of course, Decca and has also now been expanded to include Navstar GPS. In addition, it is sophisticated enough to automatically select the system which gave the most accurate position fix in that specific area.

To develop hybrid GPS receivers was therefore almost a pre-ordained course. What has been surprising though is the diversity of both the equipment able to

The GPS receiver   35

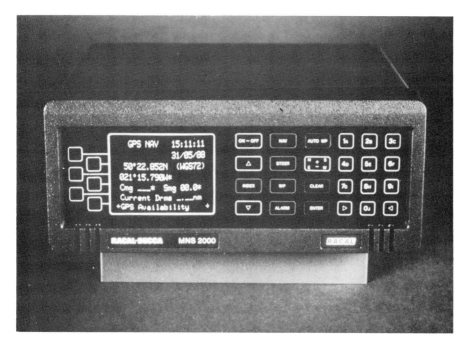

Fig. 23. MNS 2000G display

operate in this mode and the systems they are capable of tracking. Some examples are given below of hybrid receivers are given below :

Table 2. Hybrid GPS receivers

   Trimble 10X - Navstar/Loran C
   Magnavox 1107 - Navstar/Transit
   Shipmate RS4000C - Navstar/Decca

Another interesting development is the research being undertaken to produce a composite Navstar and Glonass receiver. A receiver that can process the information from both systems to produce a single fix will dramatically improve system performance globally, especially in the short term. In addition, it will help to improve user confidence in the final systems with inbuilt back-up at the system level. The term *redundancy* is often used for this. There are, however, many political considerations behind such a match and it may take some time to reach a stage allowing such integrated operation of the two systems. Current co-operation in this field between the two countries looks favourable with substantial agreements already being made on the inter-operability of Loran C and Chayka (the Russian Equivalent). Some limited discussions have taken place about the possibility of similar steps being taken on the GPS systems. In anticipation of

36 GPS and the ship

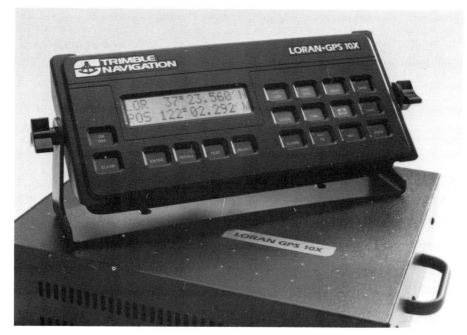

Fig. 24. An integrated GPS/Loran C receiver (courtesy of Trimble Navigation Ltd)

agreements being made, both academic and commercial based initiatives are already underway on the development of suitable hardware.

Although hybrid receivers are an important product there are a few disadvantages that need to be considered as well. This follows the well-known adage of "putting all your eggs in one basket". Should a common element to the receiver fail, for example the power supply unit or the display, then obviously all electronic navigation may be lost. If, however, separate receivers were in use, for say GPS and Decca, then complete unit and navigation back-up would be available. In many instances vessels will probably already have existing navigation receivers so the purchase of a hybrid GPS receiver may not be necessary, unless of course even more navigational back-up is required by a cautious operator.

2.3 A practical guide to purchase

As can be seen there are very significant variations in GPS hardware, but on entering the market place to purchase, the buyer will find the number of manufacturers now offering low cost GPS receivers is even more bewildering. It is, therefore, a worthwhile exercise to list the type of features that will be required for your specific operations. The selection between sequencing receivers and multi-channel is reasonably straight-forward with only high accuracy operations, such as dredging and survey, requiring the multi-channel units. It is probably a good idea, regardless of accuracy requirements, to opt for at least a dual channel receiver. The second channel is used in down-loading the satellite information (ephemeris) and allows for much quicker acquisition of new satellites. Single channel receivers tend to slow down positioning, if only for a few minutes, whilst they down-load ephemeris.

2.3.1 Navigation features

The type of navigation utilities offered by a receiver is obviously of great importance to the professional navigator. Most receivers have the facility for inputting way points and providing navigation along the great circle . Usually distance-to-go and time-to-go features are included, as are course made good and speed made good calculations. Further features such as more sophisticated track guidance options are also offered, with helmsman's displays giving visual real-time navigation data. Consideration should be given to whether a gyro or speed log input is required. These features are especially useful for more accurate dead-reckoning calculations. They could be important requirements considering the reduced coverage until the early 1990s. The ability to manually enter this information would also be of value for dead-reckoning purposes.

The level of useful navigation information is often dictated by its ease of access and the quality of display. This point will be expanded on shortly. However, it is important that the navigation fix/position information is quick to access and easy to see, preferably with some quality indicator alongside. Ideally the position display should be the default display, i.e. the display page normally on view. On certain GPS receivers this is not actually the case.

2.3.2 Operating Modes

This is a rather unspecific title to cover a rather important subject. The ability to provide the receiver with either height and/or clock information can substantially improve the current coverage and will continue to do so even after the full constellation is up and running. Of significance to the mariner are the advantages gained from the ability to operate only in two dimensions. The calculation of height is rarely ever needed on a vessel. Not only does this increase the current working coverage by allowing operation with only three visible satellites, but it also actually provides more consistent accuracies. Although GPS is designed to give three dimensional position information the satellite configuration often cannot provide the best geometry to give this to the same accuracy as two dimensional positions.

To allow operation with only three satellites the operator must be able to input into the receiver the height of his antenna above sea-level. This should be measured as accurately as possible, although within a metre or so is usually adequate for navigation purposes, especially as changes in vessel loadings and tidal effects will always introduce variations. Once given this figure, the receiver should automatically be able to correct for the global variations between Mean Sea Level and the vertical reference of the WGS84 satellite system (for a fuller explanation see Chapter 5). This can change by as much as 100 metres over the world's surface. The means by which this error is removed is by knowledge of the

38 GPS and the ship

Geoid/Spheroid separation figures. It is important that the receiver is capable of correcting for this and that this specific point is checked with the receiver manufacturer.

A further way of increasing coverage hours, at least for the next year, is to be able to operate on two satellites only. To achieve this an external clock input must be provided for the receiver. Some low cost navigation receivers do actually have this feature. This clock must be accurate to a degree only attained by atomic clocks (such as rubidium or caesium atomic frequency standards). As these are likely to cost more than the receiver this is not considered to be of much real significance to the marine navigator. More information on this technique, however, can be found in Chapter 5, The GPS Detail.

2.3.3 User interface

This is the rather overly technical term used for what actually refers to the operation and button pushing part of the receiver's use. Surprisingly, this should be quite a significant element in the selection of a receiver. With the major advances made in single chip computers it is quite impressive to see the amount of user functions that can now be crammed into a small receiver. It can also be quite confusing.

Figure 25 gives an example of a menu structure, the standard way of giving access to different levels of information in a GPS receiver. These menu structures tend to get quite complicated and often quite time consuming in recovering all the information you might need. This, under certain conditions, could be at the least awkward, at the worst dangerous.

Before buying a GPS receiver try and get a test-drive for at least an hour or so but if possible, preferably for a few days. A menu structure that looks at first sight to be very complicated may actually prove to be quite user-friendly after a little while. Alternatively you might find it very clumsy to get the type of information you want quickly. Some receivers the authors have tested appear to have been designed with no feeling for the operator at all, especially one who has other tasks to worry about on the bridge.

Another very important part of the user interface is the screen of the GPS receiver. There seems to be a race to design the smallest GPS receiver possible at the moment, with little thought to the fact that the display quality and information is actually critical to successful use. If the receiver is to be permanently installed on a vessel's bridge, size may not be that critical but visibility will.

As a result of this trend the use of Liquid Crystal Displays (LCD) is becoming very common. These are very good value and have a very low power consumption but are rather difficult to see, unless the user is directly in front of the screen. In direct sunlight they can be almost unreadable. A new development known as a Gas Plasma Display moves away from some of these limitations.

The best displays use light emitting diodes (LED) or a standard small television screen (CRT—cathode ray tube). LED displays give good readability, also have low power consumption but normally only give a few lines of information at once, which can be somewhat limiting. CRT's give the best of all displays, sometimes with graphics, but these are less common and may be more expensive. They also

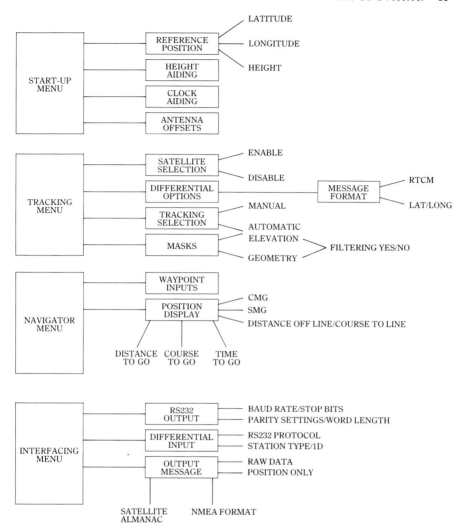

Fig. 25. A generic receiver menu design

have a higher power rating and tend to be bulkier. It may well be that it is intended to integrate the GPS receiver with a track plotter/navigation computer, in which case the receiver display quality will be of less significance.

2.3.4 Hardware interfacing

The hardware interface is the means by which data can be passed from the GPS receiver to another device such as a printer or track plotter. It is through the integration of the precise positioning information, with additional data such as in

40   *GPS and the ship*

electronic charting, that the full potential of GPS is realised. The type of applications open to such integration are reported at length in this book.

Unfortunately, although the potential is great in this field, the reality is quite different and specific integrations are only likely to be realised with comparably significant investment, unless the equipment has already been interfaced to the relevant peripheral, e.g. track plotter or PC.

The receiver manufacturers have generally standardised on the structure of the data outputs, with most utilising the RS232C scheme. This is a way of coding the data so that it can be read universally and is based on serial data transmission. The term *protocol* is used to describe the characteristics of a serial data string. Serial refers to the fact that the data is passed out sequentially bit by bit as opposed to parallel techniques, which pass out 8 bits of data at a time down an eight wire cable. Some manufacturers have adopted a slightly different scheme called RS422, but in most cases provide the option of a converter capable of producing RS232C as well.

Although the data protocol is standardised the message contents often are not. Each receiver, with a few exceptions, offers its own message types designed to provide the information that specific manufacturers consider necessary to provide quality controlled position information. Often receivers give you the option of outputting different types of messages, adding further to the difficulties of interfacing.

Attempts at standardisation have occurred with the introduction of the NMEA 0180 and 0183 formats, available for example in the Magnavox 5400 GPS receiver. These tend to provide rather limited information for specific purposes such as inputs to autopilots, but are certainly an important option.

2.3.5 Hardware integration

The ability to integrate a GPS receiver with say an electronic charting computer will depend on the policy of the provider of the charting package. Often companies will provide both products, in which case you will be limited purely to their GPS hardware, such as with the Shipmate RS5310 GPS receiver and their RS2500 colour plotter. This makes for easier maintenance agreements and guaranteed compatibility.

If, however, the charting software comes from an independent software house then there may be greater flexibility in terms of software customisation, often with the software house indicating which GPS receivers can be interfaced. Alternatively, as the purchaser is probably making a significant investment in the charting package, then it is quite normal to make it a proviso of the purchase that the package is specifically interfaced to the relevant hardware. If it is an intention to integrate the GPS equipment with additional computers, or even just to operate in a differential environment, this area of hardware compatibility becomes quite critical. Any receiver purchase should be carefully considered and discussions undertaken with all equipment providers in the integrated package. This area of compatibility between technologies is repeatedly one of the most time consuming and expensive elements to any sophisticated navigation package.

The GPS receiver 41

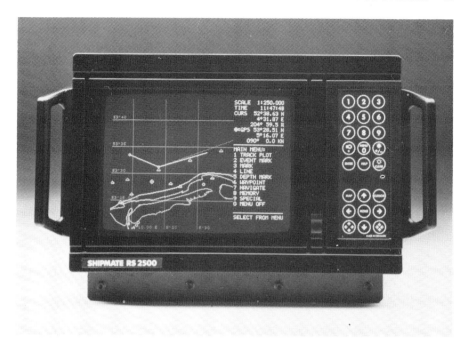

Fig. 26. Shipmate RS2500 colour plotter

2.3.6 Power supplies

This aspect of hardware is quickly becoming less of a problem with substantial development work being directed at producing lightweight and flexible AC and DC power units. Many marine GPS receivers are designed to take in DC voltages between 10 and 28 volts unregulated. In addition some can also take in AC inputs between 100 volts and 250 volts. This, of course, should be checked to ensure compatibility with the vessel's power supply. It is interesting to note, though, that of all the GPS hardware failures experienced by the authors over the years many have been traced to the power supply units. Vessels are often notorious for the vagaries of their power supply and this may well be the root cause of the problem. The possibility and practicality of carrying replacement power supply units which are easy to install may, therefore, be a worthwhile consideration, as is, probably, a comprehensive list of board spares.

2.3.7 Antenna installations and cabling

Another area in which grief has been caused time and again in the authors' GPS operations has been that of antenna installations and cabling. This may have often been caused due to the fact that most of these installations were of a temporary nature, but even so it does highlight a weak spot in GPS equipment design. The frequency that GPS operates on is not particularly suitable to transmission down long lengths of cabling nor to processing within modern receiver architecture. In response to this the manufacturers often amplify the frequency in the antenna and then downconvert it to a more suitable frequency for sending down cables. They then may further change the frequency by mixing it in the receiver to ease the signal processing functions. This is detailed at greater length in the receiver design section in The GPS Detail.

The result of these techniques, or their non-implementation, is that there is often a restriction to the lengths of cables and the type of cables that can be used for GPS antenna installations. In some cases long cable runs of greater than 30 metres (100 feet) need special low-loss cable or amplifiers. This is usually because there is no down conversion of the signal in the antenna. It is also wise not to have too many connectors in the cable as these provide weak spots, often subsequently found to be the cause of signal grounding and loss. For exceptionally long runs of 60 metres or more it may be necessary to use the special armour shielded low-loss cable, although this is notoriously difficult to run and very expensive. All this might sound overly detailed, but poor cable connections and the use of incorrect cable types for the length of runs can be a serious problem if advice is not taken. As the GPS signals are very susceptible to signal shielding, multi-path and satcomms interference, the best installations are often on points high and remote from the bridge often requiring substantial cable runs. Receivers providing both preamplification and down conversion of the signal in the antenna do not suffer as much from cable restrictions.

Interference from radars and some satellite communications systems is a serious problem for the marine GPS operator. The location of the antenna near to either of these installations and specifically in the beam width of the transmissions could, at the least, interfere with the reception and, at the worst, burn out the antenna and damage the receiver. The separations necessary for successful operation will vary from receiver to receiver but this should be checked. The more expensive receivers invariably have much more sophisticated RF design specifically to limit interference damage.

2.3.8 Differential features

This subject is covered in much more detail in a later section, which is essential reading for anyone interested in differential operations. However, it is important to note that differential capabilities are becoming part of the internal capability of some marine GPS receivers such as the Magnavox 4400 GPS receiver. It is also important to realise that these are being standardised around a recommended format. This format is very rigid and allows successful differential operation with some ease. It is referred to as the RTCM 104 format. For differential operations to become a widespread and reliable operation for precise navigation in, say, congested waters and port approaches, standardisation is essential.

The GPS receiver 43

Fig. 27. Magnavox MX 4400 GPS positioning and navigation system

2.3.9 Service agreements and software upgrades

Servicing and maintenance agreements are an important part of the purchase of any significant piece of electronic hardware such as a radar or navigation receiver. Although the projected cost of a marine GPS receiver is relatively low, circa $US 5000, this is a field of technology that is changing rapidly. Questions regarding forward product developments should be asked of a preferred receiver manufacturer. Related to this is the question of product compatibility to these future developments. Already manufacturers have abandoned product lines in GPS hardware, leaving users with obsolete receivers with limited spares support. With the current speed of change it is quite likely that this will happen again. Although no manufacturer is likely to admit to such a negative feature, ask questions regarding the evolution of their current product and their future plans for new models or upgrades.

Although the hardware is an important element in forward product policy, software is also critical and, in fact, a lot easier for the designer to change. Software upgrade costs should also, therefore, be raised with the manufacturer. Normally such upgrades are secured with an additional percentage charge above the basic receiver cost. Again, to justify the payment of such a charge discuss the types of improvements the manufacturer intends making, at least over the period of the software maintenance agreement.

Obviously certain types of software changes should not be paid for, in particular software bugs. Even in the most reputable of GPS receivers currently on the

44 GPS and the ship

market bugs of both a major and minor nature have been discovered. In fact it is often only after significant field use that these types of errors are ironed out.

A further type of software modification that should also be resolved by the manufacturer free of charge is that those enforced on them by changes in the Control or Space Segment of the system by the American DOD. Already three such significant changes have occurred, requiring the manufactures to modify their software. One was with the datum change from WGS72 to WGS84 in January 1987. A second change came in describing the age of satellite ephemeris implemented in November 1988 and thirdly some slight format changes occurred in the signals of the first Block 2 satellite launched in March 1989. At least two of these changes resulted in many commercial receivers failing to operate at all. The manufacturer's attitude as to whose responsibility such occurrences are should be discussed.

CHAPTER 3

Differential GPS

1. The differential concept

1.1 Introduction

The term Differential GPS has already been discussed in this book as a means to provide a higher level of GPS performance and accuracy. This chapter will concentrate in more detail on the value of differential operation to many levels of users.

In many senses this chapter may be more at home in the section "The GPS Detail", but due to the significance of this subject at a practical level it is included in the main body of the book. The authors also believe that operation in differential mode has a much greater significance than just improving accuracy. Fundamentally it provides confidence in the system, a critical point when viewed in the light of the general concern about the accountability of the GPS systems to the average user. Differential GPS allows the operating authority to stamp their own personality on the system and enforce their own levels of quality control and repeatability.

The concept of differential GPS, commonly referred to by the acronym DGPS, is not new and has been applied to many navigation systems in the past. It relies on the assumption that certain types of errors, which can degrade the performance and accuracy of a system, are common to all users (within a given range). If these errors can be calculated at a point their application to the data of other users, as a correction, will cause them to be removed or reduced. More succinctly Differential GPS involves the removal of correlated systematic error between a reference receiver and a remote user. This is, of course, GPS jargon.

Obviously, the main assumption behind differential techniques is that they improve the overall system performance. This unfortunately may not always be the case. With the presence of still undefined means of accuracy denial in the hands of the US military, to be introduced in times of alert, there are no guarantees. However, the differential technique inherently allows these types of problems to be at the least made apparent. This area of debate is actually very important to the DGPS question and will be covered more fully.

Assuming that differential techniques are able to remove the effects of selective availability, as is currently the case, then practical accuracies at the five to ten metre level are now achievable as opposed to the stand-alone figures of one hundred metres (95% confidence level).

46 Differential GPS

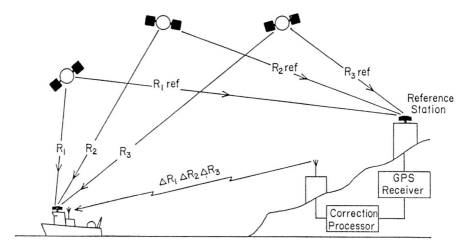

R_1 corr = $R_1 + \Delta R_1$ etc.

Fig. 28. Differential GPS

There are other limiting factors to the differential concept that need to be appreciated. Yet the implementation of a differential system can still be undertaken in the realm of the informed system user. This is probably why this subject has dominated the GPS literature for the past five years, especially since the announcement of selective availability.

It is the authors' intention that, in this section, enough information will be provided for the reader to define his own requirements and if necessary enough detail to implement such a system.

1.2 System design

To be able to define errors in any navigation system the correct value of the observation either must be known or be calculable. The errors inherent in GPS are visible to the user only as an error in his position or position uncertainty. These can only be quantified if the user actually knows where he really is. Prior knowledge of his position will allow these errors to be identified and also allow them to be refined and possibly removed.

It is, therefore, apparent that in a differential system one receiver/user must know where he is, i.e. be at a known reference point. This obviously has an immediate and major cost significance. Someone must be willing to operate a receiver installation at a static, unchanging point. This moves the system into the area of the service provider and thus there needs to be a requirement for such an implementation, be it for capital gain or safety reasons. Certainly, up to date, all differential services available are offered in the commercial environment.

For the errors to be calculated correctly this reference receiver must know its position in the same framework as the satellites. In other words the same tape

The differential concept 47

measure must be used at both the reference station and the mobile, with the same units being read. In GPS terms this tape measure and units are the WGS84 spheroid and datum, discussed more fully in Chapter 5, Section 3.5.

Once these errors have been determined, if they are to be of any value, they must be available to another user as a correction to improve his positioning accuracy. For navigation this must obviously be in real time and quickly enough for the information not to become stale. For differential GPS to be successful it is reasonable to expect this to be achieved in under fifteen seconds, although again this will be controlled absolutely by the form of selective availability adopted.

Table 3. Differential error budget
Under selective availability

Error source	Stand-alone	Differential
Space segment		
Clock instability	15.0 m	0 m
Ephemeris errors	40.0 m	0 m
Orbit errors	5.0 m	0 m
User segment		
Ionospheric delays	12.0 m	1.0 m
Tropospheric delays	3.0 m	0.5 m
Multi-path	2.0 m	2.0 m
Receiver noise	2.0 m	2.8 m
Total root sum squared	44.8 m	3.6 m

The medium by which the corrections to these errors can be made available to users is another complicating factor in the differential system design. In a practical, automated system it would require the use of a suitable radio data link. The subject of data links is quite a complex one but as it is a critical element to real time differential operation it will also be covered in some detail.

The ionospheric and tropospheric components of the differential error budget are probably the most interesting. Certainly, delay induced by the ionosphere is correlated, as in real terms the signal path will differ little when seen against the height of the orbiting satellites. About half of the ionospheric error is reduced through the use of a special model actually transmitted in the satellite data message. Dual frequency operation would allow its virtual removal. Differential techniques certainly help to reduce the residual error, but this will become less succesful as range increases between the reference and mobile. Maximum workable separations are in the order of 1000 kilometres.

Tropospheric delay is a much smaller error and for many applications is of less interest. For the highest accuracies, though, even this is of concern. Again this is controlled by distance separation and, most importantly, the presence of a weather front between the reference and mobile. Input of meteorological information may help this.

SAME SATELLITES MUST BE USED BY BOTH REFERENCE AND MOBILE

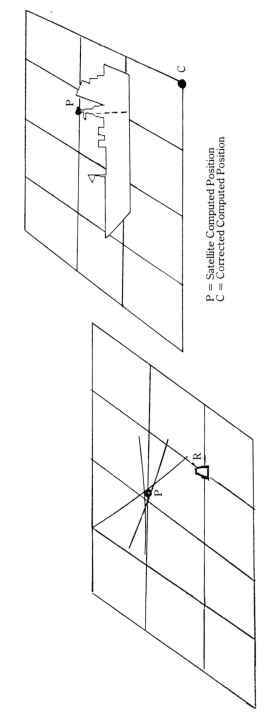

P = Satellite Computed Position
C = Corrected Computed Position

R = Reference Position
P = Satellite Computed Position

Fig. 29. Common view (block shift) correction technique

1.3 Techniques of correction

There are two ways of defining and applying differential corrections in DGPS. The first and simplest to implement is through the use of position corrections. A reference receiver at a known point calculates a position from a given combination of satellites. This is then compared to the known position and corrections in terms of delta latitude, longitude and, if relevant, height are computed or, more correctly, delta X,Y,Z the co-ordinate reference frame of the satellites.

The second and more sophisticated method of correction is to determine corrections to the actual range measured to each satellite. These are known as delta pseudo-ranges or pseudo-range corrections. It involves more complication than position corrections, but does offer greater flexibility to the user. With this method the reference receiver must calculate the true range to each satellite from his known point, i.e. the straight line distance from his cartesian position (X,Y,Z) to the satellites' cartesian position (X1,Y1,Z1). This is then known as the computed range and from this is subtracted the actual receiver measured pseudo-range to give rise to the correction. This technique has gained the most favour and is certainly the most flexible.

The pseudo-range is contaminated by a clock bias and the handling of this bias is one of the more difficult aspects of differential pseudo-range systems. In fact, as differential systems are trying to remove common errors, they must successfully resolve the clock error before the correction is passed. These clock errors are not related between two separate receivers, although they could all be, theoretically, handled in the computation to position at the mobile.

It is apparent that the pseudo-range correction procedure is the more complicated of the two techniques, with current trends being to incorporate this into the more sophisticated receivers. Alternatively, commercially available software packages operated on small PCs (Personal Computers) can be bought to achieve the same end. These software packages can cost from anything between $US 5,000 and $US 15,000.

Advantages and disadvantages

The major advantage of the pseudo-range differential method is that the mobile user is totally independent in his selection of satellites to compute to position, assuming that the reference station is transmitting corrections for all satellites in view. With a completed constellation, over eight satellites will be available for use at any one time. A simple marine receiver will probably only use up to four of these. For corrections to be calculated to all visible satellites this would require a receiver with up to ten channels at the shore station.

In a position correction system both the reference receiver and the mobile user would have to adopt a common view schedule where each would be observing the same satellites at the same time. With current receiver design, this common view schedule could only be guaranteed if the receivers were operated manually and told which satellites to select by the operator. Position correction systems might also suffer to a greater degree from the effects of selective availability than pseudo-range systems. Here the error features on all satellites are combined to produce a less easily controlled error environment.

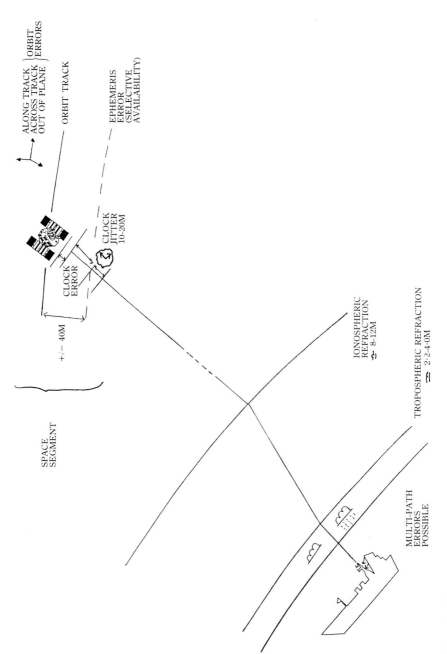

Fig. 30. Pseudo-range error sources

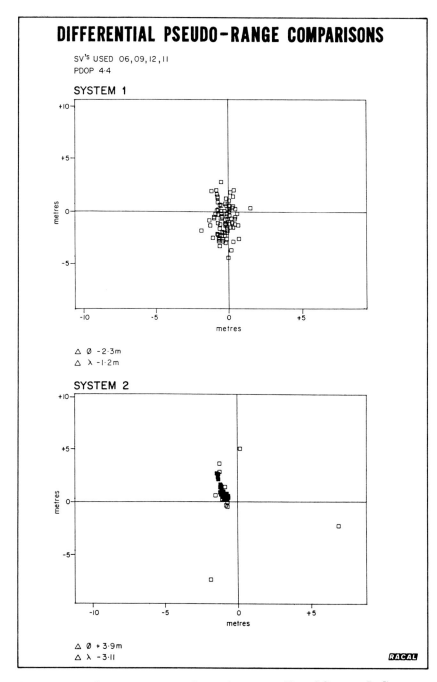

Fig. 31. Pseudo-range comparisons (courtesy Racal Survey Ltd)

52 *Differential GPS*

Pseudo-range corrections also have the advantage of being valid for greater distances than position corrections. This is because position corrections are dependent to a degree on the geometry of the satellites being used in the computation of position. The greater the distance between the reference station and the user, the less similar the relative geometries of the two receivers and the less accurate the corrections. These errors become significant beyond about five hundred kilometre separations.

However, as most differential operations for marine applications will tend to be at shorter ranges and close to shore, this distance factor may be of little significance. For the marine applications that would probably require differential techniques, such as port and harbour approaches, automatic docking and dredging applications, cost and simplicity will probably be just as important considerations.

An important technical point regarding pseudo-range differential systems is that the corrections are currently much less transportable than position corrections. What this means is that pseudo-range corrections tend to be very receiver or software dependent and that they often only are accurate if the same receiver/software that derived them applies them. This results from the difficulties in resolving the clock errors already mentioned and from a significant degree of unique processing that is undertaken in the receivers. Position corrections, however, are, hopefully, identical between different makes of receivers.

This has quite a significant implication to, say, a Port Authority which will have no control over the type of receivers installed on vessels using the port, but which for safety reasons might want to operate within a differential positioning framework. In addition, it is becoming apparent that differential pseudo-range services would incur a much heavier installation cost, requiring ten channel reference receivers, possibly additional software and some significant expertise.

One major weakness of position correction systems, especially if safety is of primary concern, is that the reference receiver may track a satellite that the mobile cannot obtain, due possibly to signal blockage or low signal levels. This invalidates the position corrections and, as such, is not a suitable approach for general navigation where safety and confidence are paramount. This would be less of a problem to a dedicated port service possibly only used for harbour surveys and the like. Here the mobile user could enforce his selection on the reference station, if necessary, and incur significant cost savings over a pseudo-range system.

1.4 Pseudolites

Pseudolites are an interesting concept related to the subject of differential GPS, but also related in another way to the satellites themselves. The term pseudolite refers to the use of a ground based transmitter which exhibits exactly the same signal characteristics of the GPS satellites, in fact a pseudo-satellite. Not only does it provide an additional ground-based range to work with, but it also acts as a differential monitor making observations to the spaceborne satellites and deriving corrections in the required format. This information is then modulated onto the look-alike GPS transmission in the same way as the satellite navigation data.

Pseudolites are still only currently conceptual with no units commercially available. However, some receivers, such as the Magnavox 5400, do already have the facility to incorporate such transmissions. This is achieved through utilising the satellite codes between 25 and 32, currently not allocated for true production satellites. Pseudolites do have a significant application for the airborne use of GPS, especially in runway approaches. There are many hurdles to be crossed before such an application could be realised. Other potential applications could be for high accuracy port and harbour control work where a number of such units could be used to provide an independent positioning service, less dependent on the space vehicles themselves. This will obviously only be realistic if the cost considerations are not high and mutual interference not a problem.

Pseudolites are, by design, only line-of-sight transmissions and, if their use becomes likely, care will be needed in their implementation. As such, certain technical recommendations have been made by the Radio Technical Committee for Marine Users, who have the task of defining DGPS specifications. These are designed to limit the possibility of separate pseudolites operating on the same code from interfering with each other:
1. Maximum separation from User 50 kilometres (27 nautical miles)
2. Minimum separation from another pseudolite 54 kilometres (29 nautical miles)

Pseudolites apart from being valuable differential tools, in that corrections could be transmitted to the user as part of the GPS signals, also give the significant advantage of improving the local area satellite geometry. This will still be of significance in the completed constellation when there will still be periodic instances of poor geometry. Pseudolites are unlikely to have an impact on the reference station/monitoring side of GPS for some years hence. Certain high precision applications, such as automatic docking operations, might hasten their introduction.

1.5 Differential GPS and selective availability

The requirement for selective availability has already been discussed in the opening chapters of this book. Yet, over the years, differential GPS has always been put forward as a possible means to remove these errors and recover the full accuracy potential of the system. There are still no guarantees even though at the time of publication we have been exposed to the realities of this degradation. This is because there are apparently many levels of accuracy degradation that the Americans (or Russians) could introduce, geared potentially to the severity of any perceived military threat. The ability of differential GPS to remove, or satisfactorily reduce, the effects of Selective Availability (SA) depends on two major factors:
1. The type of SA
2. The rate of SA

In practice certain types of SA may not be removable by differential GPS, such as the introduction of a high rate clock jitter onto the code transmissions, thereby reducing the chip resolution. This is more unlikely as it will degrade all users including favoured ones. As such, this would truly be accuracy denial. Other forms of unrecoverable SA can and have been (at length!) postulated.

Currently the status is that the SA being operated on the Block 2 satellites can be removed by differential techniques, but only if operated at a sufficient update rate.

54 Differential GPS

This point becomes critical in the design of a differential service. It is not so much the size of the errors that is of concern, but how quickly they change. The profile of the current SA is of an ephemeris error (epsilon error) of about forty to fifty metres on each satellite, which does not change significantly between hourly navigation data updates. Of more concern is a clock dither term which gives an additional error growth of about 0.1 metre per second, but showing change in rate direction at approximately three minute intervals. At this direction change the error would need the most rapid correction.

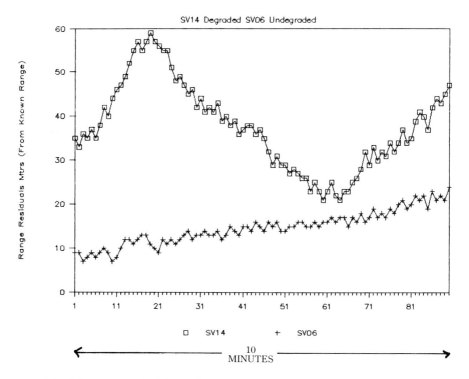

Fig. 32. Selective availability plot

If a differential service wants to maintain a range accuracy of better than five metres it will, therefore, need to pass corrections more frequently than every twenty seconds. To achieve a system where SA error is restricted to within the smoothed pseudo-range noise figures, updates at better than twelve seconds will be required. Obviously, for marketing and commercial reasons, differential service providers may well use this update figure as an important trump card. But this all needs to be put into the context of the realistic accuracies of a differential system, of between five and ten metres. The success of a differential GPS system in the environment of Selective Availability will, therefore, depend to a significant degree on the speed that the corrections can be made available to the vessel. This is a function of the capability of the data link used to transmit the information. In coarse terms higher speed data links are easier and cheaper to implement over shorter distances. This is in the favour of the maritime industry where most applications would tend to be closer to shore. An exception to this rule is in the proving ground of offshore oil exploration where the requirements for high accuracy positioning seem to be daily moving further off the continental shelf.

1.6 The Radio Technical Committee for Marine Services

If a differential mode of operation is to be adopted by the marine navigation community then it is necessary for a common world-wide standard to be used. A ship must be able to decode and apply differential transmissions in any part of the world, be it a port approach such as Europoort or a congested seaway such as the St. Lawrence. Some concern has already been voiced in the real compatibility of existing hardware. This is not intended to deter, but just to highlight areas that still need attention. However it is critical, at the least, that there should be standardisation of formats.

This requirement was first investigated in June 1983 when a workshop held at the Transportation Systems Centre in the USA developed some general recommendations regarding the data formats and computations behind Differential GPS. Subsequently, the Institute of Navigation requested that the Radio Technical Committee for Marine Services (RTCM) further investigate and return standards for a differential broadcast service. In November 1983 a special committee, 104, was established containing many of the industry's leading names. Three working groups were set up; one to investigate the actual message content, the second to investigate the communication requirement and the third to investigate the pseudolite concept.

These studies resulted in a series of standards being published in November 1987, which were subsequently revised in March 1988 and February 1989. The findings have been well-documented (ref.) and offer a very well-thought out quality control orientated format. Its success is mirrored by the degree of institutional acceptance already gained in the USA and world-wide. Receiver manufacturers have already started to incorporate the necessary features into the receiver hardware to allow hands-off differential computations to be undertaken directly within the units. If DGPS is to become a reality then these developments must continue and in such a way so as not to enforce specific hardware limitations on the maritime community. It is therefore essential that the representative agents of the world's maritime community further reinforce the need for standardisation. In practice these must go beyond formats to the specifics of the actual calculations and further must include the Glonass navigation service.

The RTCM message format

The actual format for the passing of differential corrections, at least in terms of its structure, have been modelled very closely on the navigation message of the GPS satellites. This offered advantages in the case of pseudolite operation, but would also allow the use of special mathematical techniques, by necessity already

56 Differential GPS

resident in the receivers. The format includes a series of message types to cover many different aspects of GPS operations above and beyond differential requirements.

For normal differential operation six different message types are of significance. The standard message, known as Type 1, gives the basic information necessary for successful operation. This includes, primarily, the accurately time-tagged satellite ID number (PRN) and the range correction, in metres. It also includes a figure describing the rate of change of this correction in metres per second. This allows the mobile receiver to predict corrections should a message or two to be missed. A figure estimating the performance of the satellite, as observed by the reference station, is also passed. This is called the User Differential Range Error (UDRE) and indicates in metres the stability of the satellite range.

Another important indicator is also passed in the Type 1 message. This is the health of the satellite again as seen by the control segment. The final element to the Type 1 message is the issue number of the satellite ephemeris. The ephemeris is the data used to compute the position of the satellite and, for the range correction to be meaningful to the user, he must be using the same issue.

As can be seen, this message provides additional information above the range correction to protect the user from other possible errors. For example, the use of an

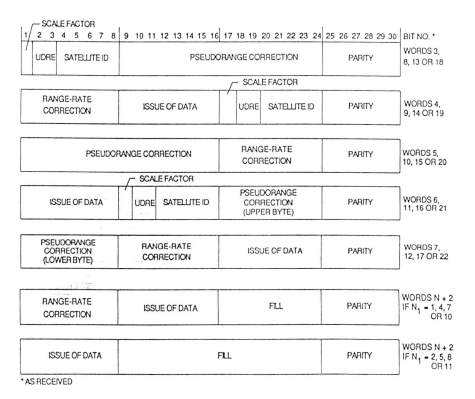

Fig. 33. Type 1 message format

unhealthy satellite or an unstable satellite. It is these features of quality control that make the RTCM recommendations so valuable.

Table 4. RTCM message types

Type Number	Title
1	Standard differential corrections
2	Delta differential corrections
3	Reference station information
4	Carrier surveying information
5	Constellation health
6	Null frame (test message)
7	Marine radio beacon almanacs
8	Pseudo-lite almanacs
9	High rate differential
10	P-CODE differential
11	L1/L2 delta corrections (C/A Code)
12	Pseudolite station parameters
13	Ground transmitter parameters
14	Surveying auxilliary message
15	Ionosphere/troposphere message
16	Special text message
17	Ephemeris almanac
60–63	Differential Loran C corrections

The other message types recommended as a minimum for marine differential operation are built on this premise of quality control. The Type 2 message is an additional correction to be used in conjunction with the Type 1 should the mobile user be operating on an older ephemeris issue than that used to derive the corrections. A Type 3 message provides information to the user on the status of the reference station, whereas a Type 5 message details further the health of the satellites as analysed at the reference station.

One final message type of special significance, we believe, to successful and reliable operation is the Type 16 message. This is a free field text message that will allow special messages, down-time warnings or alerts to be passed to the mobile user. This has been used numerous times in our own experience of differential operation to advise of potentially serious difficulties and provides an essential human touch to the system. This feature should be incorporated into all differential systems and preferably be alarmed in the mobile units.

Data rate

When data is passed down a cable it can be passed in two ways, either using parallel techniques or by serial. Parallel data transmission sends the required bits of information down up to seven lines at the same time, a different bit down each line. Serial data transmission, most commonly used in data links, assumes the bits of information are being sent sequentially down one line. Serial data has a protocol all of its own, involving such factors as the speed at which the data is passed, measured in the bits of information passed a second (the baud rate), characters which indicate when a new block of data is being sent or when a block is finished (start and stop bits), and how many bits of information should be combined together to make a sensible piece of information, i.e. a word (word length). A bit of information is either a binary 0 or 1, usually eight bits making a word. The arrangement of 0s and 1s in a specific order have a sense, exactly like dots and dashes in morse code. The most common arrangement scheme is known as ASCII. The RTCM select committee suggested that a data link operating in a differential service should be capable of passing data at speed of fifty bits per second (50 baud). This again conforms to the data rate of the navigation message superimposed onto the satellite signals. Fifty baud is a fairly fast rate for data to be passed over a data link and should handle many possible variants of selective availability.

It is important to realise that 50 baud refers to the actual throughput of data. Many radio links need error correction to remove the corrupting effects of interference and so the same message may be sent a number of times to guarantee its reception. Although these messages may have a baud rate of 50, if they have to be sent five times to guarantee their uncorrupted reception then the actual throughput rate is only 10 baud. The RTCM format has some limited error detection capability built in, but the need for additional coding is subject to the data link frequencies adopted and their susceptibility to radio interference.

1.7 GPS integrity monitoring and quality control

GPS integrity monitoring is an area of quality control for GPS operators that is gaining in popularity even though, in the guise discussed, it does not currently exist. Integrity monitoring refers to a function similar to differential GPS, but less sophisticated. In essence it is envisaged as an independent service which monitors the performance of the satellites, continually broadcasting information indicating whether the system is operating satisfactorily or not. This, it is hoped, will give the user more confidence in using the system and will quickly be able to highlight any failings. In reality, this is already part of the control segments responsibility. In the case of severe problems a satellite is marked unhealthy and subsequently ignored by receivers. The main problem appears to lie in that certain flaws in the system are not tagged in real time and civilian users are discovering the problems themselves. This might be relatively easy to achieve in a gross sense or if another high accuracy navigation system is available for comparison, but under normal conditions some types of problems might go unnoticed with potentially disastrous results.

This type of problem tends to be highlighted at the moment, probably because the system is still in a test and development phase. In the future, when the system is being heavily utilised both by the military and civilian community, it is likely that suspect satellites will be flagged in real time by the control segment. If the confidence could be built into the integral monitoring of the system then possibly independent monitoring may not be required.

Regardless of these points there still appears to be quite a lobby developing for an independent integrity monitoring service, especially for groups like the FAA (Federal Aviation Authority), where confidence and reliability are critical to safety and any system errors need to be discovered in seconds. This is perfectly understandable and any service which can improve reliability and safety must be discussed seriously. However, as in differential GPS, there is still an overriding difficulty—the transmission medium for such a service.

For the maritime community, integrity monitoring, unlike a differential GPS capability, is a feature that will probably be required for the complete voyage. This is especially so if confidence in the existing control of the system is not increased. For this to be practical it would be best implemented over a satellite communication link such as on Standard C. Here the delay time limiting the suitability of the medium to differential GPS is not so important. In addition, differential GPS is more geographically limited than an integrity monitoring service. For integrity monitoring to be successfully implemented over satellites, all GPS satellites visible in the communications satellites footprint must be considered, with some significant overlap. The value of integrity monitoring services in a more local area must also be given some thought. Again, like DGPS, the choice and availability of a suitable data link becomes a problem, especially now if there is increased frequency competition caused by differential requirements! But in this scenario differential GPS itself might still provide a better confidence builder than an integrity monitoring service. For example, the geographical restrictions of differential GPS are not so apparent over a local area as they are, say, over an ocean-wide area. If a transmission link is required for a monitoring service, why not put it to more use by actually allowing any errors to be potentially recovered through differential techniques. Differential GPS is, in itself, the ultimate integrity monitoring service. In fact, it is the authors' opinion that this is just as important a role as improving accuracy. The RTCM 104 differential format, in fact, includes significant monitoring features and quality control information in its own right.

Unfortunately, though, it is not all black and white. The data rate overheads for a differential link are likely to be more than a simple monitoring service, therefore requiring a wider bandwidth, more difficult to obtain at least for the civilian sector. In addition, a differential service requires the mobile user to be more sophisticated with in-built differential capability in the receiver. The two approaches need careful consideration by the relevant advisory bodies. Certainly, with the reductions in equipment costs, an integrated differential capability should not really incur any significant cost disadvantage. In fact, it seems sensible that this should be made an integral feature of all marine GPS receivers, even if it is not used. It may well be that the best solution is a compromise. Integrity monitoring providing the necessary quality control and confidence in the GPS system for the ocean phase of a voyage, whereas differential GPS provides both higher accuracy and more importantly error recovery capabilities for continental shelf or congested channel areas.

2 Data links

Introduction

The selection of a suitable data link is certainly one of the more difficult practical decisions facing the differential service provider. This is especially so if the system is for general use, such as port approach navigation. This section is designed specifically with the port and harbour authority in mind, users for whom differential GPS has a practical meaning and who might be interested in setting up their own service. This is generally going to be for more specific applications controlled closely by the harbour authority, such as dredging and harbour surveys. In local areas a simple differential service will be a lot cheaper to install and run than, say, a dedicated microwave positioning chain.

Wider area differential services are more likely to be operated either completely in the commercial sector or by official or regulatory bodies such as the coastguard authorities, or even government transport departments. In the case of the USA, the coastguard appear to be making the running in differential circles whereas in the UK studies have been sponsored by such bodies as the GLA (General Lighthouse Authorities). A data link encompasses the modulation of the differential correction data onto a suitable carrier frequency which can be successfully and simply decoded by the mobile user. Certainly advances in low cost data link technology has eased this task, but it is still important that standardised formats and technology are adopted at all stages.

Error detection and error correction are also important considerations in defining the link protocol. Error detection really just refers to the inherent ability to define that an error has occurred in the transmitted message. This is usually achieved through parity checking. If only 0s and 1s are being sent then it is fairly simple to add a few extra bits to make a message add up to an even number or an odd number. If a bit is decoded wrongly then the message will not add up to the expected solution, i.e. odd or even. Error correction requires much more sophistication and allows some degree of contamination to be reversed. Error correction techniques can become quite complicated and are generally outside the capability of the smaller system designer.

It should be realised that simple data links can literally be bought off the shelf and even these often contain some error detection features. Normally they are transparent to the type of data being sent. This means that they do not need to understand the message or data being provided and they just pass it through as given, often adding error detection characters, then removing them at the receiver before passing the message onward.

2.1 The compromise

In the selection of a suitable radio frequency for a data link there has to be, unfortunately, a series of major compromises—range versus performance versus simplicity versus expense. In the sphere of marine navigation it is unlikely that differential GPS will be required at great distances offshore. A general rule of thumb is that precise navigation or increased confidence will be more necessary nearer to shore than further away. It would be very difficult, for example, except through the use of direct broadcast satellites, to pass information at ranges in excess of 1000 kilometres at the necessary update rates. This is not too much of a concern as at these ranges the differential information itself starts to become invalid.

In any differential service there are two points of view to consider, that of the provider of the service and that of the user. In terms of technology the system needs to be geared to the lowest common denominator, i.e., the user. DGPS will only be successful on a broader, multi-user base if the required communications technology is cheap and widely available. In addition, the services need to be compatible between different user areas. DGPS is by no means a necessity for the marine navigation community, but its advantages, specifically in confidence levels and error recovery, must not be outweighed by its costs.

2.2 The differential options

There is a whole radio spectrum to choose from when considering the setting up of a DGPS data link. However, certain practical and cost features need to be realised. The overriding concern is frequency allocation. In many parts of the world the radio bands are heavily congested and issue of new frequencies, especially for data transmissions, is a rare occurrence. Economic re-use of available frequencies is one approach, as is the modulation of the data on existing carrier transmissions. Certain radio wavelengths, as will be detailed, are also significantly easier to obtain permission to use. In general terms the longer the radio wave, the further it travels. The shorter the wave, the faster data can be modulated onto it. This again is the compromise of range versus performance versus cost. Short range high frequency data links are significantly cheaper and easier to install than the longer range links.

Long wave/low frequency 30 kHz to 300 kHz

This radio band is probably a non-starter in the differential stakes as a general medium. Although it provides the longest ranges, it also requires the heaviest investment unless use is made of existing transmitters. Tall masts and high power requirements characterise low frequency transmissions. The band is also congested and it is unlikely that enough space could be found, free of strong daytime or nightime interference, for a fast enough and reliable link service. It is also not a frequency band particularly suitable for world-wide or even region wide adoption as relatively small changes in frequency require substantial changes in user equipment. However, some differential services already exist in this frequency band. These have been superimposed on to an existing navigation service, but with no change in signal characteristics. A system known as Pulse/8 is operated in the Euroshelf area, the South China Sea and the Yellow Sea by Racal Survey Ltd.

The system itself operates at 100kHz and is similar in design to Loran C, but with more sophisticated monitor control allowing it to achieve accuracies of better than 30 metres. Although the data rates are limited by the need to retain the same signal characteristics, RTCM type 1 message updates can still be achieved at under twenty seconds for all visible satellites. This system has the advantages of already being in situ, having long range capability (in excess of 700 kilometres) and being inherently an independent position service. Further development, in terms of updates rates is still ongoing to operate the system in an environment of full selective availability.

Fig. 34. 100 kHz transmitter

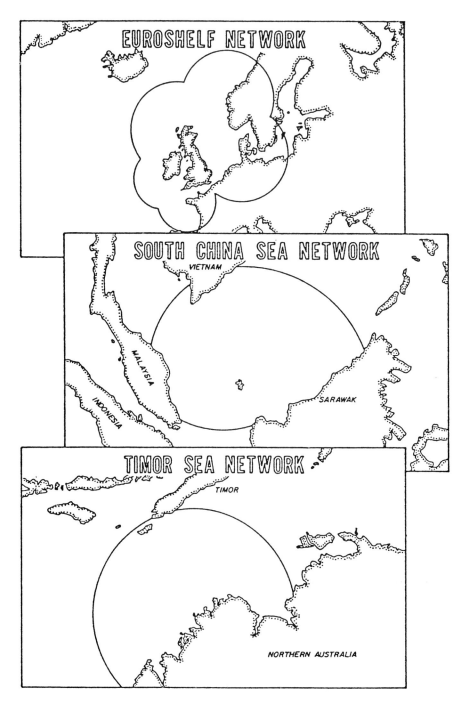

Fig. 35. Existing differential coverage provided to the oil exploration industry (courtesy Racal Survey Ltd)

Medium frequency 300 kHz to 3 MHz

This frequency band probably offers the best possibilities for a wider area, multi-user DGPS service by integrating longer range capabilities with lower installation costs, ship and shoreside. Medium frequency actually covers the transition between the groundwave propagation of the long wave frequencies to the skywave propagation of the high frequencies. As such, the lower frequencies in this band can offer, with higher power, ranges up to 500 kilometres, whereas lower power transmitters and higher frequencies can still provide reliable information up to 200 kilometres.

The main problem regarding the MF band is, again, frequency allocation with the developed world having very strict regulations regarding new transmissions in this range. However, for the professional mariner, the MF band has one very important saving grace, the Marine Radio Beacon (MRB). The potential of using the marine radio beacon has already been assessed by most maritime organisations around the world. In fact, if the ongoing trials prove satisfactory then it may well be in this guise that the professional mariner will first come across DGPS services. Already the United States Coast Guard has invested significantly in investigating this option. Some embryonic systems are already in operation.

Substantial studies already undertaken in this field have concentrated on discovering safe ways of modulating the necessary data onto the transmissions without affecting the existing purpose of the transmitters [Per K. Enge et al.; "*Coverage of a Radio-Beacon Differential Network*", Navigation, Vol. 34, 1987]. Marine radio beacons operate in the frequency range of approximately 285 kHz through to 325 kHz. They actually cross, somewhat, the boundary between the LF and MF bands. Currently the most favoured way of incorporating the differential data is to pass it over a sub-carrier of the main frequency. This means using a frequency centered very close to the main transmission, but not far enough away to be distinguished by other rf users as a separate transmission. Normally an offset of between 325 and 500 Hz from the main transmission will suffice.

There are many advantages to such a methodology primarily through the use of *in situ*, well located, beacons already covering the areas of heavier coastal traffic and port approaches, as for example, around the US eastern seaboard. Being *in situ* means minimum costs are incurred in initiating such a service and, of course, frequency allocation is not necessarily a problem. Another major advantage is that the beacons are already operated by official organisations who are committed to providing reliable and safe navigation. With standardisation to RTCM recommendations and with user equipment relatively cheap at this band, a truly compatible, global DGPS service could be offered.

In some parts of the world MRBs will unfortunately not provide the complete solution. In Europe, for example, it will require the successful co-operation of many different national operating authorities to implement such a service. Secondly, MRBs may not cover all areas where differential services will be required. But thirdly, and most importantly, in some locations MRBs will not be suitable hosts due to existing time-share features. Rolf Johansson (ref. *International future navigation needs: Options and concerns*, Navigation Vol. 34 no. 4 1987–88) noted that in North West Europe, for example, a whole chain of MRBs operates on a single frequency, namely 308 kHz. They do so on a time sharing basis with each

Fig. 36. A medium frequency differential GPS station (courtesy Differential Technology Ltd)

66 Differential GPS

beacon only transmitting every six minutes for a period of one minute. This would be a completely unworkable situation for a DGPS network working under selective availability.

High frequency 3 MHz to 25 MHz

This frequency band offers an important alternative to the Marine Radio Beacon concept, possibly as a means to fill in the holes. Again, HF offers distinctly different transmission characteristics dependent on which end of the band is used. At the lower end ranges, with the necessary power, of a few hundred kilometres can be achieved reliably. In this context wider area GPS coverage could still be offered.

The major advantage of HF is the relatively low cost of setting up such a system from scratch and the higher chance of frequency allocation being approved. There are significantly less strictures placed on this band, but this does mean interference problems are less well controlled. Techniques such as frequency multiplexing (the use of say three separate frequencies) could protect against this. Again the problem in sophisticated solutions is the ability for users to access that technology.

Very high frequency and ultra high frequency 30 MHz to 300 MHz

VHF and UHF transmitters offer a very significant tool to the differential operator. This technology is cheap and simple to install, providing very high data rates over local areas, within line of sight. For most port and harbour applications this is probably ideal. Such equipment can be bought off-the-shelf and often simply plugged into the differential system, being transparent to the data being sent. Low power requirements and transportability make it very suitable for low level users. Frequency clearance is amongst the easiest to obtain, usually with no problems.

The satellite dimension

The use of communication satellites for the passing of differential GPS corrections has always had strong promise. In some instances this has even be realised. An example is the Star-Fix service operated by John Chance Associates in the Gulf of Mexico (see Chapter 1 Section 2). Here, an integral, independent satellite positioning service is incorporated with differential GPS transmissions at a high update rate. Such systems could be seen to provide the ultimate redundancy.

Most satellite communication systems operate on the C, L- or Ku band with frequencies in the L-band ranging between the 12–14 GHz spectrum, for example. There are already, world-wide, many licensed operators of such satellites who are now able to transmit data and even sell channel space to third parties. Initially most communication satellites were licensed to individual cartels who were very restrictive about such practices. Satellite communications, at first glance, offer the ideal solution to differential GPS requirements, but there are unfortunately practical and cost considerations that will restrict their utility to many potential users. The renting of a suitable data channel for dedicated differential use may cost in excess of $US 100,000 per year. This immediately limits such operations to larger commercial concerns or state controlled bodies. A more practical consideration is

Fig. 37/38. Diffcell—a combined GPS receiver and VHF transmitter (courtesy Measurement Devices Ltd)

68 Differential GPS

that for most satellites services, data has to be passed to an uplink station operated by the system controllers. As such, the differential monitor site would either need to be co-sited at the uplink station, or a sophisticated and fast data transfer network would have to be designed to get the data to the site. Most off-the-shelf mobile satellite terminals now available are designed to work with part of a time-sharing, packet service, and would not provide compatibility with dedicated channel availability.

The most serious problems associated with satellite communications are the possible delays inherent in many of the existing or projected services. Although the data rates may be exceptionally high in comparison to, say, an MF data link, there is often an uplink delay or queue system of, sometimes, in excess of twenty seconds before data can be messaged out to users. These problems tend to occur when a differential service is being implemented in an environment designed for normal data or telex transmission. This may not be the case in dedicated satellite link options.

CHAPTER 4

GPS: applications and implications

Introduction

It is difficult to separate the implications from the applications when dealing with the introduction and the subsequent effects of the GPS technologies on the existing maritime industry. This section is arranged to give an overview of the types of changes to current navigational practices which may be accelerated, instigated or curtailed by the introduction of GPS. Specific examples have been selected for a more detailed analysis and have been included in later sections.

1. GPS and coastal navigation

Although GPS will undoubtedly have many applications for deep sea navigation, particularly in terms of fuel efficiency and safety, it is probably in coastal navigation where it will have the greatest number of users. The subject of deep sea navigation and position monitoring is dealt with more fully in Chapter 3, Section 3, Position and Data Reporting, whereas the following chapters will deal more with the requirements for precise navigation in coastal areas and congested seaways. A coastal passage is generally the most dangerous portion of any vessel's journey and the vast majority of marine casualties occur during it. The reasons are straightforward, first, an increased risk of collision due to increased traffic densities, often compounded by geographical bottlenecks such as the English Channel; secondly, an increased risk of grounding either on the coastline or off-lying obstructions. Because of these facts the vast majority of both conventional aids to navigation and electronic position fixing services cover these coastal regions. When the impact of GPS is discussed it is very often in the context of its competition with conventional positioning systems, such as Loran C, Decca and Omega. However, the purpose of this chapter is to discuss the potential impact of GPS on all aspects of ship navigation, not just on what positioning sensor will be carried.

Coastal navigation, a definition

The geographical limits to coastal navigation, i.e. the transition point between oceanic and coastal passage is a product of two different groups of considerations.

70 *GPS: applications and implications*

1 Physical factors

These include the physical geography of the coastline and the seaward extent of navigational hazards such as reefs, shoals, oil platforms and prevailing weather conditions.

2 Technological factors

These include the size of the vessel, particularly its draught. Pre-1945 vessel draughts would have rarely exceeded 10m, today 500,000 tonne tankers with draughts of 25–30 m regularly ply the world's major shipping routes.

The extent of the ship's horizon is another important consideration. Historically the ship's horizon was determined by how far the lookout could see from his vantage point and the prevailing visibility. With the advent of Radar a modern vessel's horizon extends beyond the visible horizon. The range of this electronic horizon is determined by the height of the radar scanner, just as the extent of the visible horizon is dependent on the height of the observer above the sea surface.

Vessels' information gathering capabilities have changed dramatically since 1945. Electronic navigation systems, Racon, echo sounders, radio beacons and radio lighthouses all provide 24 hour navigation information, most in all weathers. Whereas at one time the physical proximity of a coastline would have determined

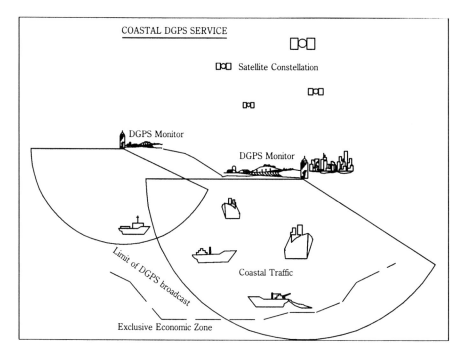

Fig. 39. Coastal DGPS service

whether a vessel was undertaking oceanic or coastal passage, it is now increasingly the technological factors which determine the transition. Many vessels would consider themselves coasting when they start to pick up Decca coverage, or, perhaps in the future, when they can receive shore based differential GPS broadcasts.

The 200 nautical mile Exclusive Economic Zone (EEZ) makes a convenient geographical boundary between oceanic and coastal navigation and is the one which we shall adopt in this book. A vessel entering a country's EEZ may well face restrictions on fishing, exploration, whether commercial or scientific, dumping of spoil or waste and general anti-pollution controls. Around the North Atlantic the 200 nautical mile limit also marks the approximate geographical extent of the continental shelf, as well as the seaward limit of many conventional navigation systems and, indeed, possibly differential GPS coverage. Perhaps most importantly, the volume of traffic increases significantly within the EEZ and, therefore, the risk of collision multiplies.

1.1 Conventional electronic positioning services

There are currently two distinct levels of medium-to-long range permanent positioning services to be found in coastal areas. The first level, or Group One, may be described as low-cost, low-accuracy and includes systems such as Decca and Loran C. The list of users include merchantmen, fishermen, aggregate dredgers and pleasure craft. The second level, Group Two, can be described as high cost-high accuracy and includes precise survey navigation systems such as Pulse/8 and Hyper-fix. These are used by a much smaller number of specialized clients, in particular the offshore oil industry.

The comparison is drawn between these two groups and GPS to highlight the fact that high accuracy positioning services have often been available in areas of high traffic density, but have not been widely used for reasons of cost. They also outline the types of service environments that more advanced GPS users may work within. For example, during the last ten years the European continental shelf from the Bay of Biscay and Western Approaches through the North Sea have all been covered, continuously, to an accuracy of better than thirty metres. Yet this has not made the transition through to the marine community in general. There is obviously a trade-off somewhere down the line of accuracy against cost.

When trying to assess the impact of GPS and, particularly, differential GPS in coastal areas, a good place to start is to compare and contrast the existing levels of service to GPS and the differences they offer their respective users. GPS effectively crosses the boundary between these two, once distinct, levels in both cost and accuracy. This must result in its widespread adoption and possibly introduce applications previously not even considered.

1.1.1 User access

The lower cost Group One navigation services are available to an unlimited number of users fitted with the appropriate receiver. Access is unlimited and achieved purely through purchase of the hardware. Access to the second group is

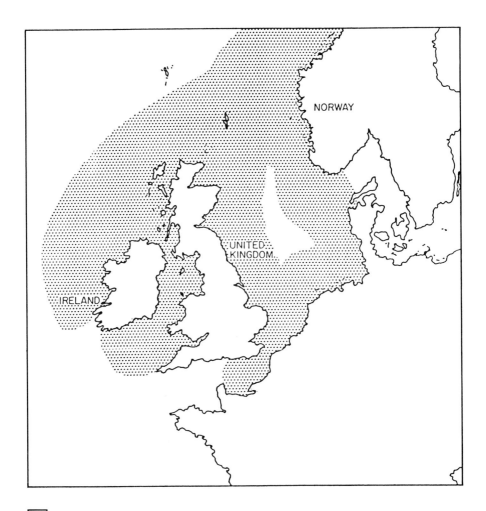

Existing High Accuracy Positioning Coverage (using 2MHz Radio Positioning Systems)

Fig. 40. Precise positioning in the North Sea better than 10m

usually by arrangement. For example to gain access to Pulse/8, a high accuracy Loran C type system, a user must hire a receiver from one of the licenced operating companies. Group Two systems might also have a physical limit on the number of users that can be accommodated.

Access to GPS is somewhat less distinct. Certainly access to the standard 100 metre positioning service of the Navstar system will be straight-forward and unregulated. Use of a higher accuracy differential service will be dependent on the ability to decode the differential data transmissions. This may be operated in the service provider environment of the Group Two systems or may be provided free of charge by regulatory or safety bodies such as in the Group One systems.

Access to the Glonass system is currently a bit more difficult to presuppose. At present it seems a standard service will also be offered at the 100 metre level, comparable to Navstar GPS. This assumes the political situation between East and West continues to make steady progress. A missing element in Glonass operation at the moment is a commercial Glonass receiver. Certainly those exhibited by the USSR would be unlikely to be adopted for mass production. A Glonass receiver or even a hybrid Navstar/Glonass receiver will probably be manufactured by arrangement in the West, although this is difficult to see within the strictures of the transfer of technology limitations imposed by Congress.

1.1.2 User cost

Decca and Loran C are free to use, although indirect payment may be made through buoyage and light dues. They are certainly not free to operate and maintain. Receivers are available from a number of different manufacturers with mass production and a competitive market, realising low prices in relation to the second group. Receiver costs may be as low as $US 1000 in total. To use a system from the second group usually entails hiring a receiver and the signals from the service providers. Costs may be as high as $US 1000 per day and rates are generally calculated for non-continuous operation.

Standard service GPS user costs will also be limited to the purchase of the relevant hardware, although for many years user charges were considered. The impractability of levying such charges eventually quashed these ideas, although this cannot still be taken for granted. As the technology of the GPS receiver is significantly more advanced than a Decca or Loran C unit prices are going to be somewhat higher. Receivers on average will cost under $US 5000. This should further reduce because of the mass market potential of GPS, attracting many manufacturers and the economies of scale. This is clearly illustrated by the Japanese interests with over fifteen receivers under development. Prior to GPS, relatively little interest was apparent from this country regarding navigation equipment. The major GPS market for such companies will actually be land navigation as opposed to marine.

The higher accuracy differential GPS services may well be maintained again through buoyage or light dues as in the Group One systems. Alternatively, they may be charged as a service by Group Two providers, possibly on a pay-as-used basis, or through levies placed on the data decoding equipment. User costs also have a more practical realisation in the era of GPS. Most existing navigation systems are regional in nature and a vessel embarking on a long oceanic voyage may require a panoply of navigation equipment. For example, a voyage from Rotterdam through the Panama Canal would pass through Decca coverage (Europoort, Western Approaches), require mid-ocean Transit positioning and then Loran C through the Caribbean to the Panama Canal. This would be further aided by Radar and possibly ARPA, all in all a very expensive set of equipment to buy and maintain. GPS fundamentally breaks down these boundaries, with the same system and the same accuracies available world-wide. The possible rationalisation of bridge equipment with the introduction of GPS is discussed shortly.

1.1.3 Operator costs

Operator costs are likely to be the main cause for a substantial change in the navigation infra-structure in association with the introduction of GPS. Group One systems such as Loran C and Decca are very expensive to maintain and operate. In fact, the US Coast Guard initiative of transferring chain control to local national interests is spurred on by this consideration. In addition, the recent discussions instigated by the British Department of Transport regarding the future of Decca with respect to wider scale introduction of Loran C are also a function of cost considerations. In Group Two systems the real cost of maintaining the shore stations is passed directly on to the user, resulting in significantly higher user costs.

The GPS alternative will incur little cost to civilian bodies involved in the provision of navigation services. Concerns about reliability and redundancy will be more easily assuaged if Glonass and Navstar operation can be combined. Confidence can be further increased by offering differential services. Recovery of many satellite error conditions can be simply achieved within a differential framework and it allows real-time system monitoring. The cost of maintaining and operating a medium-to-long range differential station from a single transmitter, will be substantially less than maintaining a whole chain of radio positioning stations, with similar coverage. It may well happen, though, if national bodies do start to provide services based around the GPS technology, that charges may be levied at this level by the American and Soviet governments.

1.1.4 Accountability

The first group of systems are owned and operated by governmental or quasi-governmental organisations, whether civilian or military, and are maintained from the public purse. The second group are in general owned by private companies and operated for corporate profit. In both instances accountability is high, with the primary customer being the marine operator. GPS is, by design, a military system and, although the operation of Transit has proven that this can be compatible with civilian operation, it does have implications to the individual countries supporting navigation services. Although the average user may be less concerned about the systems' origins, it might have political considerations at a higher level. A total reliance on navigation from either GPS system obviously places European or Third World countries heavily at the bequest of the two Super Powers. This is unlikely to happen completely and this point will probably guarantee the existence of at least one of the regional navigation systems.

1.2 Traffic management and separation schemes

Safe navigation in congested or coastal waters is now often more about the management of vessel movement and the maintenance of safe distances between vessels. In fact, the emphasis on safe navigation is increasingly about transferring decisions on strategic ship routing from the bridge to the shore. Both traffic separation schemes and general coastal navigation management are becoming areas of increasing importance and interest. A less obvious implication of accurate global positioning may be in its effects on maritime legislation and standardisation. If a majority of vessels are known to operate a common system with common accuracies, then it will be much easier to define a common infrastructure for navigation procedures and possibly even wider ranging maritime legislation.

The first traffic separation scheme was introduced in the Dover Strait in September 1967. The first implementations were based on proposals made by the institutes of navigation in France, West Germany and the UK and were introduced on a voluntary basis. The initial objective for establishing traffic separation schemes was to significantly reduce the number of collisions between vessels by separating opposing streams of traffic. Since the introduction of the new collision regulations in 1977 it has been compulsory to comply with the requirements relating to traffic separation schemes laid down. In the same year the IMCO sub-committee on the safety of navigation extended the objectives of such schemes to include the following:

1. The reduction of the incidence of head-on collision by separating opposing streams of traffic
2. The reduction of the danger of collision between crossing traffic and shipping following the traffic lanes
3. To simplify the patterns of traffic flow in congested areas
4. To organise the flow of traffic in areas of intensive off-shore activity
5. To organise traffic flow to avoid areas where vessel movement is dangerous or undesirable
6. To provide guidance to vessels with regard to areas where water depths are critical or uncertain
7. To guide traffic clear of fishing grounds or organise traffic through them

Cockroft (1983) has shown that the introduction of traffic separation schemes did indeed result in a considerable reduction in the incidence of collision between vessels proceeding in opposite directions. "The reduction", he observed, ". . . has been most noticeable off north-west Europe and applies, almost exclusively, to the incidence of collisions in restricted visibility . . .."

It is interesting to compare the apparent success of traffic separation schemes in preventing collisions, to the study carried out by Lithgart and Wepster (1985) into the impact of the introduction of navigation systems on the incidence of groundings. They concluded that the gradual widespread introduction of such systems, particularly in N.W. Europe and North America, had not significantly reduced the yearly percentage of ships lost due to grounding accidents. Both the *Torrey Canyon* and *Amoco Cadiz*, two of the biggest pollution incidents in N.W. Europe, were due to grounding, not collision. Most recently the *Exxon Valdez* incident off Alaska has highlighted this point. It is this fact which has extended the role of traffic management and separation schemes to include coastal protection and highlighted the need for improvements to be made in legislation and the automated navigational requirements of large vessels.

When considering the possible route of a traffic separation scheme, particularly when it is for coastal protection or to avoid offshore installations, the planners are faced with a dilemma. On the one hand they may well wish to route the traffic as far away from the coastline or offshore installations as possible. On the other, it must be possible for ships within the scheme to fix their position. The problem is most

76 GPS: applications and implications

acute for regions without any long range electronic navigation coverage, yet adjacent to major shipping lanes, or those traversing ecologically sensitive areas. At the moment they are generally restricted to introducing such schemes within radar range of land.

The worldwide implementation of GPS, and the requirement for large vessels to carry an electronic positioning system, will undoubtedly benefit such areas. If, as has been suggested, precise navigation could be incorporated in real-time to accurate nautical charts then concerns about grounding could be more easily assuaged. Vessels' draft information could be incorporated intelligently into the Electronic Chart Display System (ECDIS) and audibly alarmed should safety be compromised. Such a system could only be realised fully by utilising an all-weather standardised global navigation system. Terrestrial, regional services will never cover all areas of marine passage in coastal areas and the variations in type, accuracy and all-weather reliability makes standardisation difficult.

1.2.1 The specification of traffic separation schemes

The widths of traffic separation schemes could also be related to the positioning quality available and the introduction of GPS and DGPS will have an obvious

Fig. 41. NavGraphic II—an ECDIS using published nautical charts (courtesy Trimble Navigation Ltd)

impact on this as well. As section 6.12 of the General Provisions of Ship's Routing states "The minimum widths of traffic lanes and of traffic separation zones should be related to the accuracy of the available position-fixing methods." Van Riet, Kaspers and Buis in their 1985 study of a proposed 22 metre deep draught route in the English Channel, defined eight criteria which could be used to decide on the optimum route. These were:
1. Minimum water depth
2. Minimum reliability of the position fixing system
3. Minimum safety margin
4. Minimum required route width
5. Minimum required track length
6. Maximum acceptable change in course
7. Acceptability of tidal sailing
8. International criteria

Of these criteria, 2, 3, 4, and 5 are all functions of the positioning systems used. The authors recommended that navigation should be through the use of two positioning systems. If, for example, a 200 metres accuracy were specified the primary system should have this accuracy with a 99.7% probability level. The secondary system would be used to verify the primary system and replace it in case of breakdown. A navigation accuracy at this level, with a 68% probability was established as adequate for the secondary system. In the event, radar was chosen as the primary system and Decca as the secondary. The problem with using radar as the primary aid is that its accuracy is a function of range. The standard deviation of the bearing being $d/65$ and the standard deviation of the distance being $d/100$, where d is the distance in nautical miles. Decca in good coverage offers 40–50 metres at 68%. However in the English Channel there is only two lane fixing. A coastal DGPS offering 10–15 metre accuracy at 95% probability would be a significant improvement and allow the realisation of these recommendations. In addition, the constant velocity readings available from GPS could be used to improve track keeping and, in particular, measure cross-track motion.

Reliable and accurate positioning increases in importance with the size of the vessel, whether planning a route or navigating one. Minimizing the required route width will reduce the cost of any dredging or obstruction removal, which may be necessary for the safe passage of deep draught vessels. From the navigators point of view, accurate positioning will obviously improve track guidance. It will also allow the navigator to recover his intended course quickly and accurately should he be forced to deviate from it by reason of collision avoidance. This could further aid safety.

1.3 Conventional aids to navigation

Conventional navigation aids tend to be used by the smaller merchant vessel or pleasure craft. In general terms these navigation aids have suffered the most in recent rationalisations. There certainly seems to be a move towards a reliance on electronic navigation aids, which will be further compounded by the full availability of GPS. With the all-weather, global availability of GPS there may well be a move by the navigation authorities to make the smaller vessel more

78 *GPS: applications and implications*

responsible for his own safe navigation through the use of low cost GPS receivers. Conventional aids can be divided into five main sources, and in practice, incur some significant cost to maintain. Conventional aids are used most widely in defining channel approaches or hazardous features. They are often the only navigational aid available in third world ports or outside EMPF coverage. The following list details the main sources:
1. Light sources, such as light-vessels, lighthouses and lanbys (Large Automated Navigation Buoys)
2. Sound sources such as horns, bells and whistles
3. Daymarks, both fixed and floating
4. Radio aids, which include marine radiobeacons as well as VHF lighthouses
5. Radar aids, which vary from fitting a radar reflector on buoys to the sophisticated Racon

Even without the advent of GPS and DGPS, some of these conventional aids, particularly daymarks and sound sources, are in decline. Others such as VHF lighthouse and Racons are being developed and implemented more widely. Light sources can be fixed or floating and have been in use since ancient times. With the continued improvements in electronic navigation aids the need for lighthouses has diminished. Wingate (1986), cites a review of major navigation aids in the UK where no less than three lighthouses were subsequently closed. Even so, light source technology has not stood still. The automation of lighthouses and introduction of Lanbys mean that light sources can be run more efficiently. New light sources, such as metal halide, hold out the possibility of considerable savings in power consumption. Indeed the US Coast Guard hope to replace some ten thousand existing light stations with these devices suggesting some continuing reliance on this technology (Keeler 1987). The decline in light sources will still undoubtedly continue, but it is unlikely they will ever completely disappear. The most likely scenario is for the light sources which remain to be combined with another aid such as Racon or VHF transmissions or even a coastal DGPS station. Such a study has already been undertaken by the General Lighthouse Authority (GLA) in the UK.

Sound sources and daymarks are the most likely candidates for extinction. They are now used primarily by smaller craft, such as yachtsmen and inshore fishermen. Most professional mariners, be they fishermen or merchantmen, are equipped with a sufficient number of other aids (e.g. radar and Decca) so that they do not need sound sources and daymarks. Some countries have already decided that there is no future requirement for sound sources (Wingate 1985).

Marine radio beacons (MRB) are classified as long-range (100 miles plus), medium-range (25–100 miles) or short-range (less than 25 miles). They have a long history, first being introduced in the 1920s with Radio Direction Finders (RDF) and are now compulsory for vessels greater than 1600 tons. As a positioning system they are easily surpassed by radar, Decca and now GPS and their future role is currently under discussion by the IMO. One strong possibility is to use the allocated frequencies for DGPS transmissions as discussed in Part 3, perhaps still retaining the radio beacon functions as a back-up utility.

The VHF radio lighthouse is a rotating directional beacon in which the directional radiated signal pattern contains bearing information. It can be used to provide a bearing from a single station or a fix if two or more stations are within

Fig. 42. Co-sited MF DGPS station and lighthouse (courtesy Differential Technology Ltd)

range. It has the advantages of being cheap, easy to use and reliable. Because of its short range (approx 30 nautical miles) it will provide an invaluable aid to smaller inshore craft, the very users who will lose most from the disposal of the more traditional aids. It is likely that VHF lighthouses will remain a valuable navigation alternative in hazardous areas. Again, there may be a possibility of incorporating differential GPS information into these transmissions. Providing dual functions and redundancy to any fixed installation is certainly more cost effective than setting up an additional independent service.

Radar beacons (Racon) are being continually developed and improved and now constitute a major navigation aid to vessels fitted with radar. Whereas radar reflectors are passive, Racons are an active system. They respond to an interrogation by sending a coded message back on the same frequency as the interrogating radar. The radar operator can then identify the Racon from the coded echo, which appears on his radar display. Racons can be tuned to respond to long range or short range interrogations. To avoid too much Racon information swamping the operator, user-selectable Racons are being developed. These will only respond if so requested by the radar operator.

Racons were being suggested as a means to provide safe navigation around large portions of the US coastline. It is interesting to note that their widespread introduction may well be superceded by a chain of differential GPS stations. Certainly the flexibility of GPS and the added confidence given by DGPS operation already appears to be having an impact across the board in navigation.

1.4 Navigation equipment on merchant vessels

Beattie (1985) identified four levels of navigation equipment standards on merchant vessels. He described them as IMO carriage rules, and IMO Minimum operational performance standards (MOPS), Minimum technical performance standards (MTPS) and Customer/supplier arrangements. Carriage rules are those laid down by international and national bodies with regard to the carriage of certain navigation equipment. MOPS are generally recognised as being laid down by the IMO. MTPS are determined by a number of authorities, including the IEC (International Electro-Technical Commission), the ISO (International Organisation for Standardisation) and INMARSAT. Customer/supplier arrangements are the practical end of the requirements, where the theoretical standards are transformed into cost-effective, working systems.

With regard to electronic navigation aids, the growth of IMO carriage rules is a direct consequence of the SOLAS (Safety of Life at Sea) Conventions since 1914. The effect of these conventions has been a proliferation of mandatory equipment for the forty thousand plus vessels covered by the various SOLAS protocols.

The table below shows three different scenarios for navigation equipment standards on a merchant vessel engaged in international trade. The current scenario might typically apply to a vessel engaged in transatlantic trade. The vessel must comply with IMO and US Coast Guard regulations and so will have a high complement of equipment.

Table 5. Equipment rationalization with the introduction of GPS, DGPS and the electronic chart

Current	Partial Rationalization	Full Rationalization
Nautical chart	Electronic chart video	ECDIS
Nautical publications	Nautical chart plotter	
Magnetic compass	Magnetic compass	Redundant CMG from GPS*
Gyro compass	Gyro compass*	Gyro compass
Gyro repeater	Gyro repeater	Redundant
Radar	Radar*	Radar*
Second radar	Second radar	Second radar*
Radar plotting	Electronic plotting	ARPA
ARPA	ARPA	Electronic chart ARPA
MRB/DF	MRB/DF	DGPS
	DGPS	Navstar/Glonass
Echo sounder	Echo sounder*	Echo sounder*
Log	Velocity from EPFS*	Velocity from GPS*
Sextant	Sextant	Redundant
EMPFS: Primary/Secondary		
Decca/Loran C	GPS/	GPS/
Transit/Omega	Loran C/Decca	Navstar/Glonass

* = direct input into an electronic chart.
DGPS = Differential GPS
MRB/DF = Marine radio beacon/direction finder

The North Atlantic is well serviced by all the main navigation systems and the vessel is therefore likely to carry at least two types of navigation receiver. At least one of the radar sets would be fitted with ARPA, the other being used as backup or by a second navigating officer during stand-by's and periods of restricted visibility.

Other than the radar being fitted with ARPA, none of the remaining navigation aids are integrated. It is the duty officer's role to obtain information from the various aids, assimilate that information and act accordingly. Part 2 of this book deals more with the integrated electronic bridge in association with the GPS receiver. It suffices to say here that the shipping companies of the industrialised nations are moving towards rationalising and integrating the ship's bridge.

The second scenario depicts the equipment levels on the bridge after what the authors have called partial rationalisation. This scenario assumes global, two dimensional GPS coverage and the acceptance of the electronic chart as a substitute or replacement for paper publications. The number of coastal terrestrial navigation systems may well have been reduced to Decca and Loran C/Chayka. The former, perhaps, as the preferred terrestrial system in the waters of the EEC, the latter as the preferred system in North America and the Soviet Union. Vessels which have continuous coverage from at least 2 EMPFS are no longer obliged to carry a log, velocity being determined from the navigation data. Other sensors, such as gyro and echo sounder, are displayed on the electronic chart, as is a radar outline of the adjacent coastline. This scenario might depict the transitionary phase of GPS acceptance.

The final scenario depicts full rationalisation of bridge equipment and the widespread acceptance and reliance on global positioning. The object is to reduce the number of different pieces of equipment to a minimum and ease the duty officers

Fig. 43. 2690 BT ARPA on board Sealink ferry Horsa (courtesy Racal Marine Electronics Ltd)

workload. Those systems that remain have at least 100% redundancy. This scenario assumes global two dimensional positioning from at least two satellite systems. Differential GPS broadcasts have replaced all but the most specialised of terrestrial navigation systems for port approach and congested channel operations. A standard service GPS capability for general navigation is mandatory onboard merchant vessels.

The only remaining sensors are radar and echo sounder, both of which are displayed on the electronic chart. All route planning course maintenance and collision avoidance decisions are made at the electronic chart consol.

Scenario two will be technologically feasible by 1992, scenario three by about 1995. The two deciding factors will be the cost to the user and the acceptance of such equipment levels by the IMO.

1.5 Alternative GPS applications

This section will give an overview of some of the different types of applications GPS may well be utilised for, above and beyond general navigation. It does not cover specific applications such as position reporting services or port navigation which are covered in more detail in following chapters. These two applications give a suitable forum for detailing the practical and commercial factors regarding GPS implementation and offer scenarios where GPS is possibly most comfortable in its innovation.

1.5.1 GPS for merchantmen

For coastal navigation, merchant vessels currently use Omega, Transit, Loran C, Decca, radar and sextant/azimuth.

Omega is only accurate, in general terms, to within 1–2 miles and is therefore of limited use in areas where navigation is restricted, e.g. by traffic separation schemes or depth considerations. Although it has an accuracy of 0.25 nautical miles (Reit et al, 1985), Transit can only give a position on average every 90 minutes. For vessels travelling at high speeds, or requiring fixes at short intervals (e.g. 6 to 12 minutes), it is really only suitable as a check on the other positioning systems.

Sextant and/or azimuth fixing will generally not give accuracies better than those obtained by using radar, Decca or Loran C. Its main drawback is that it is time consuming, especially when frequent fixing is required and, of course, relies on good visibility. Conversely, its main advantage is that it is low tech and hence very reliable, easy to understand and easy to use.

Decca and Loran C are proven and popular coastal positioning systems. Land path, diurnal and seasonal errors are well documented and understood. They are widely used by many different groups and so receivers are competitively priced. The main disadvantage is their regional nature, meaning, for example, that a Decca receiver is required for navigation on the eastern seaboard of the North Atlantic and a Loran C receiver for the western seaboard.

The most widely used electronic positioning system for coastal navigation is radar. It offers continuous, all-weather positioning information. It can also be used world-wide in sight of land, is not under the control of any organisation or state and is unique in that it combines positioning and collision avoidance information on one screen. This is a critical function of radar that even GPS has difficulty matching. Radar techniques, such as parallel indexing, are also a useful method of ship guidance. The main disadvantages to radar positioning is that it has a limited range with respect to land visibility, and even if floating marks are used, accuracy can be degraded due to movement by tides and currents.

Of all the user groups, the introduction of GPS and DGPS may well have the most to offer to the coastal mariner. A truly global positioning system covering all the world's trade routes. With perhaps 500,000 plus users world-wide, unit costs would drop, leading to the possibility of vessels carrying two GPS receivers as well as two radars. In addition, such a large number of users would mean good world-wide support. Combined with the advances in satellite communications and the electronic chart, GPS presents what amounts to a revolution in navigation. It must again be emphasised that it is only in association with these complementary advances in marine technology that the real benefits of a global, all-weather navigation system are realised. On ocean voyages continuous accurate positioning in combination with automated ship control systems will allow the most efficient course to be maintained with optimised fuel consumption. This is a practical realisation of the just-in-time philosophy and can save many thousands of dollars per ship year.

1.5.2 GPS for fishermen

The majority of the global fishing effort is concentrated on the continental shelf and, hence, within the Economic Exclusion Zones. This also means that much of the fishing effort occurs within the theoretical range of shore based navigation systems. One area where such coverage exists is the north-east Atlantic where some 12 million tonnes are caught annually. The system used for navigation is Decca Navigator and is fitted to fishing vessels of all sizes. Repeatability is the key to the popularity of Decca amongst fishermen. This means the ability to return to the same point, e.g. a wreck or shoal, by using the pattern readings on the receiver. Since fishermen are often actually quite interested in accuracy, at least in the form of high repeatability, the potential for only 100 metres stand-alone navigation may be of less significance compared to the capabilities of differential GPS. The higher return fishing vessels have already implemented a reasonably high level of electronic assistance with fish finding sonar and electronic charts as already well-utilised examples. Obviously for deep sea fishing vessels accurate global navigation is of great interest, as are potentially inter-fleet position reporting services, allowing as many vessels to locate a good shoal in the shortest time.

The implications that GPS may have on existing positioning systems may have, in practice, a greater affect on the fisherman. The new navigation technologies are probably not far from the root cause of the current UK navigation review. The aim of the very powerful fishing lobby within the EEC is to keep Decca operable for as long as possible. The ideal long term future for Decca would be to have it adopted as the preferred terrestrial system to complement and backup GPS. There is currently, however, a dialogue on the European future of Decca and Loran C.

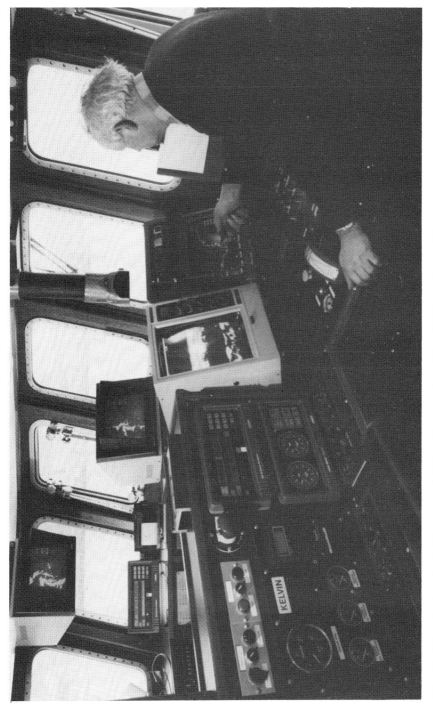

Fig. 44. The electronic fishing vessel: Wheelhouse of crabber William Henry II

Fig. 45. A seismic vessel towing a sound source

1.5.3 GPS for oilmen

Although the impact and usage of the GPS in the Offshore industry is outside the scope of this book, it would be unrealistic to discuss the implementation of GPS without reference to it. Pearson has calculated that of the estimated $US 200m spent on radio navigation systems worldwide, some $US 143m of this is directly expended by the oil companies and an additional $US 32.7m with oil company involvement (making nearly 90%). This compares to $US 9.7m (approximately 5%) for marine navigation. This oil industry expenditure is directed at a fleet of approximately 1100 vessels. The number of potential users for marine navigation systems are approximately 38,000 merchant ships greater than 500 GRT, 17,000 of less than 500 GRT, 115,000 fishing vessels and perhaps 900,000 yachts of greater than 26 feet in length (Beattie, J.H., 1985). So, although the oil industry may operate very limited numbers of vessels they do help pioneer the new technologies.

The accuracies required by the offshore industry vary from decimetres for certain offshore engineering projects, 1–5 metres for seismic exploration and pipeline inspection, to 10–30 metres for preliminary site investigations and rig moves. GPS at 100 metres is inadequate for all offshore positioning requirements. However, differential techniques are being actively investigated by the oil industry with a view to operating GPS at metric precision levels.

Differential GPS services proposed for coastal navigation will probably not match the highest accuracies required by the offshore industry however they could compete for the lower order work. High precision differential GPS services will be implemented by the offshore exploration industry, but at a technology level probably not affordable by the marine navigator, although the boundaries between the two may well become very vague. It is also important to acknowledge the fact that most differential development has been undertaken in the field of the exploration surveyor, with operational services already covering most of the Euroshelf area and the Gulf of Mexico.

1.5.4 GPS for coastguards and policemen

The application of GPS to the arena of marine policing and monitoring is certainly of major interest. Although, again, the benefits are really only harnessed when used in association with other technologies. Policing of the EEZ includes fishery protection, customs and excise functions, pollution monitoring and security of offshore oil installations. Coastguard functions include Search and Rescue (SAR), traffic monitoring and/or management schemes. The exact division of responsibility for these activities varies considerably from state to state. For example, the UK Coastguard service is concerned primarily with what we have, quite arbitrarily, called coastguard functions. The US Coast Guard has a much wider area of responsibility being much more of a policing organisation.

All the activities listed above have three common requirements, although often for very different reasons. These are currently:

Vessel surveillance

This may be by monitoring emergency radio frequencies, shore based radar installations, or observations made by surface vessels, aircraft, satellites and visual lookouts.

Vessel identification

This may be voluntary, as in the case of vessels entering a traffic management/ scheme or a vessel in distress giving its name and position. There is, however, an increasing amount of involuntary vessel identification. At the moment this usually takes the form of visual observation of a vessel's identifying marks. In the case of fisheries protection, many fishing vessels are obliged to display their identification numbers so as to be easily observable from the air. There is also the possibility of regulatory electronic tagging being introduced. This would allow the appropriate authorities to determine which vessels were in a given area when an offence, e.g. a pollution incident, occurred.

One such scheme is being proposed by the Ministry of Agriculture, Fisheries and Food (MAFF) in the UK. The requirement being proposed is for all vessels which dump spoil or sludge at sea to have on board an instrument, nicknamed the MAFF black box, which will record the position of the vessel at the time of dumping.

Vessel interception

This could range from the extremes of a rescue vessel being sent to a casualty, or a coastguard cutter being sent after a suspected drug smuggler. The essential elements are in locating and tracking the target and then passing that information to the mobile, which must then determine its own position and set the most effective interception course. The introduction of GPS and DGPS has significant implications for vessel surveillance, identification and interception. Electronic vessel tagging will require a continuous all-weather positioning system with an extensive, preferably worldwide, area of coverage. GPS fits the requirements, although there are many questions regarding the implementation of such a scheme. These include cost, reliability and the need for such a system to be absolutely tamper proof.

GPS and DGPS could influence all stages of a search and rescue operation . The initial position of the casualty is known to within 100 metres, and may well be automatically sent with the distress message. The interception procedure could also be greatly enhanced if constant and accurate position and velocity measurements are available to the mobile, possibly being automatically broadcast by the stricken vessel. Yet the adoption of GPS by search and rescue organisations must ultimately depend on its widespread use by the marine community. For example, the RNLI fits its lifeboats with Decca because the vast majority of casualties who are fitted with an electronic navigation system are fitted with Decca. This gives their service the ability to locate casualties without any ambiguity in position due to systematic differences in positioning systems or even co-ordinate reference frames. Many navigators also purely utilise the Decca lane count readings.

Although the GPS can be used for SAR purposes, it must be put into the context of other purpose-launched satellite systems such as the joint USA-USSR COSPAS/SARSAT system. This system has been pioneered by the USA, USSR, Canada and France and currently consists of two American and three Soviet satellites. These satellites orbit the earth every 101 minutes at altitudes of 750 to 1000 kilometres. The satellite orbits and footprints are so arranged that a distress signal would be received on average 62 minutes after transmission, with a maximum elapse time of 274 minutes. The satellites monitor three frequency bands – 121.5, 243 and 406 MHz, and beacons transmitting on the 406 MHz band can have their position determined to 2–5 kilometres. Other information contained on the 406 MHz band includes distinguishing whether the signal is coming from an aircraft or ship, the country of origin, the nature of the distress and the registration number of the aircraft or ship.

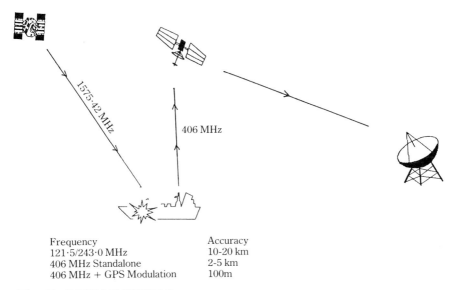

Frequency	Accuracy
121·5/243·0 MHz	10-20 km
406 MHz Standalone	2-5 km
406 MHz + GPS Modulation	100m

Fig. 46. SARSAT-COSPAS

Discussion has already been undertaken on the use of onboard GPS position information being modulated into the 406 MHz band to give even more precise position determination. Maybe this will be the arena for the first integrated Navstar/Glonass initiative.

1.5.5 GPS for yachtsmen

The single biggest marine market for GPS is the leisure market, and, in particular, the yachting market. Cheap, stand-alone GPS receivers, capable of 100 metre accuracy worldwide, are perfectly adequate for all but the most wealthy of pleasure craft owners. Unless differential capability can be introduced at a very low cost it is doubtful that the majority of yachtsmen will have any interest in a coastal DGPS service. However, if a coastal DGPS service does become widely adopted by the commercial marine community, there may well be pressure to reduce the number of conventional aids to navigation such as lightships and houses, buoys etc. This could cause considerable conflict between leisure users, who may not have any electronic navigation capability, and the authorities responsible for such conventional aids.

1.5.6 GPS for EEZ management

The management of EEZ resources is a subject fraught with conflict. For example, fishermen complain about their grounds being polluted by the dumping of waste and disrupted by aggregate dredging, cable laying and pipeline construction. The owners of cables and pipelines complain about damage by trawlers. GPS is not going to solve user conflicts. However, it does have a part to play in mapping resources and their spatial relationships. Areas such as dumping grounds or prospecting areas can be more rigorously defined with the knowledge that mariners of all types could identify their positions more accurately and using the same source. Potential transgressors would possibly lose the excuse of not knowing exactly where they were or having an incompatible positioning reference.

2. GPS for port positioning

Introduction

Today's ports are under constant pressure to provide their customers with more comprehensive and cost effective services, while at the same time responding to the navigational demands imposed by larger and larger vessels. For many estuarine ports, such as Antwerp, Rotterdam and Hull, a large proportion of their positioning activities are taken up in hydrographic surveys and the maintenance of shipping lanes and dredged channels with their associated buoyage and marks. These commonly require a dedicated network of navigation beacons providing positioning accuracies of 2 to 3 metres (Igguilden, 1986). In the case of the Humber Estuary, for example, a chain of 13 Racal Micro-Fix beacons cover some 71 kilometres of the river and its approaches. Even relatively small ports, such as the port of Blyth in Northumbria, require such dedicated networks. In this case, because of the elongated geography of the port and approaches, a network of 8 Motorola Mini-Ranger beacons are necessary.

Such networks of dedicated beacons in general meet the positioning accuracies required. However, they often present considerable logistical problems in both installation and operation and can be an additional demand on already scarce and expensive manpower resources. The use of such highly accurate positioning systems can have a considerable effect on maintenance dredging costs. Criteria established by the Atlantic Region of Public Works in Canada, for example, are in the form of a set of standard deviations of channel widths. Nominal width of a channel is determined as six times the beam of the largest vessel to use the facilities. This channel width factor can be reduced to five if a suitable harbour positioning system is used for the vessel, and to four times the beam if accurate transverse and axial velocities of the vessel can be determined. It is important to note that all these factors can be affected by cross currents of as little as two to three knots. However, this does illustrate that potential savings can be made for harbours with an extensive network of dredged channels to be maintained by using accurate positioning systems. The need for accurate positioning systems in ports and harbours is, therefore, well established and such systems are in use throughout the world. The question which we wish to address is that of the potential impact of the GPS upon the positioning requirements of such ports and harbours. To do this we will first look at how the various positioning requirements are currently met and then go on to discuss the possible use of GPS based systems to achieve the necessary requirements. This section will also detail the new technologies already being implemented in these areas which GPS complements strongly and which may initiate further advancements. GPS also crosses the navigation boundaries for ports, capable of providing the highest accuracy for surveys and dredging, but also functioning as the general navigation aid for port users.

2.1 Conventional port positioning services

2.1.1 Hydrographic surveys

One of the primary responsibilities of many ports is to provide accurate and up-to-date charts of their approaches and navigable limits. These are most often used by pilotage services provided by the port, but may also be distributed to vessels such as ferries and Ro-Ros which regularly use the port and do their own pilotage. Ports are usually legally responsible for the information contained on such charts and so it is important not only for the safety of the vessels involved, but also by reason of sensible commercial practice, that these charts are as accurate and reliable as possible. The problems of maintaining and producing such charts is often compounded by complicated environmental conditions, which lead to a continually changing seabed on timescales varying from days to years. The solution to this is either to maintain fixed channels by regular dredging or by changing the course of the shipping channels to meet the depth requirements of the vessels and sometimes a mixture of the two. Whatever solution is chosen, regular and repeated surveys are essential to maintain confidence in the published charts amongst the pilots, seafarers and shipowners.

There is a vast range of equipment available to the hydrographic surveyor to assist him in his task. However, budgetary constraints often limit his choice. The level of technology used varies widely from port to port even in a relatively small community such as the UK. The survey capabilities of most ports will, however, fall into one of three broad categories which are discussed below.

Low technology surveys

These surveys use relatively imprecise positioning methods. The methods vary from position fixing with sextants or steering between known marks within a port, to using permanent navigation systems such as Decca Mainchain or Loran C. Position fixing often involves manual plotting, with the echo sounder trace being arked at the same time.

Such low technology surveys are also usually quite labour intensive, requiring a minimum of two surveyors and a helmsman to ensure that all the data is collected at an adequate rate. For ports and harbours with only slowly changing seabed topography these surveys are often perfectly adequate and, indeed, offer some distinct advantages over the higher technology options. The positioning system requires only small capital outlay, for example, low cost yacht receivers are often used for Decca. There are also no additional costs for maintenance of a positioning beacon network. The survey methods, including quality control and interpretation of results, are well proven, and remarkably good results can be achieved by experienced personnel.

Semi-automated surveys

The level of technology required for this type of survey is well catered for by the equipment manufacturers such as Racal and Motorola and it is a popular form of survey amongst many ports. A typical survey package at this level will include a precise positioning sensor with some form of real-time track guidance and data logging. These functions are often combined into a single ruggedized unit, such as the Motorola Mini-ranger 484, Racal Microfix and Del Norte Trisponder packages.

The diagram below illustrates a typical semi-automated survey system. The operator can enter his pre-planned survey lines into the receiver unit. An online trackplot facilitates track guidance and ensures the survey area is properly covered. Fix printouts allow the surveyor to monitor the quality of the positioning and there are simultaneous marks to the echo sounder. Position information is also logged to a storage device, along with start- and end- of line times.

Accuracy requirements are usually of the order of two to three metres throughout the survey area. This requires a carefully planned network of beacons to be set up around the survey area. The beacons may be permanently installed or moved around to cover particular areas for a given period. Beacon sites must be surveyed in to a high order of accuracy and the positioning system and network properly calibrated before the desired accuracies can be attained. Careful attention must be paid to the likely operating ranges of the system and the effect this may have on signal strength and subsequent position quality. Power requirements for the beacon sites are another important consideration, as is ease of access for routine maintenance and troubleshooting.

The capital costs of such a system are relatively high, and can be equivalent to a number of years total survey budget for some smaller ports. If, in addition, the port does not have the services of a professional surveyor on the staff, the thought of planning, installing and operating such a system can be quite daunting. On the positive side, most equipment suppliers will provide advice on planning and installation of their equipment to suit the client's requirements. Most of the commonly used systems are well proven, rugged, and, if installed properly, will certainly provide the required positioning accuracy for a significant number of years. Perhaps the most important point to make is that if high accuracy surveys are required then a port currently has no choice but to buy such a system.

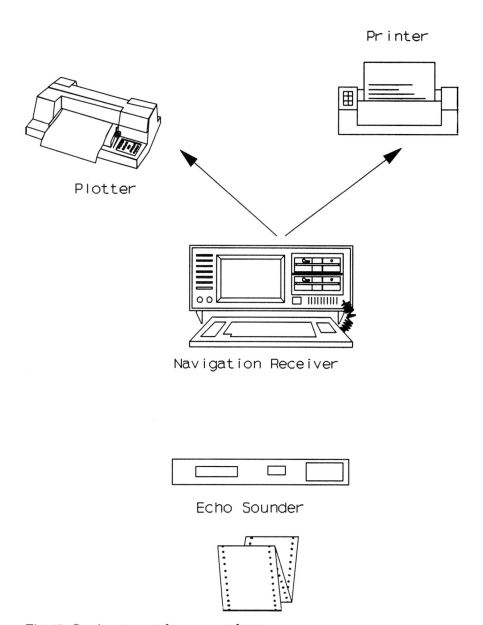

Fig. 47. Semi-automated survey package

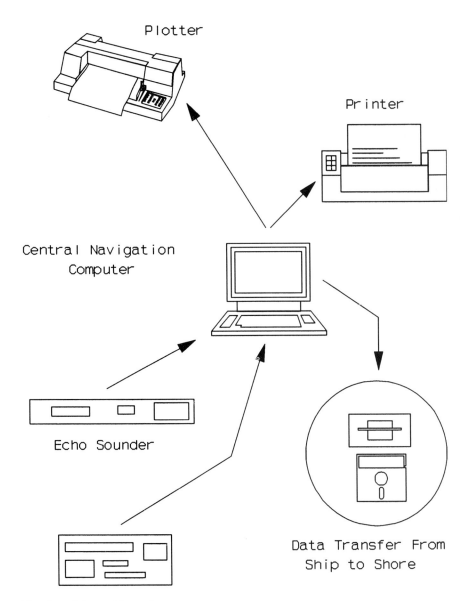

Fig. 48. Automated survey system

Fully automated surveys

The introduction of the personal computer (PC) into the hydrographic survey market has led to the evolution of the fully automated survey package. The survey methods discussed previously may allow automated post plotting of the vessel's track (from the logged data), but often still require manual interpretation and plotting of the sounding data. Once again there is nothing inherently wrong with this traditional method of producing sounding charts. Experienced personnel can produce such charts at a reasonable rate, and this method has the advantage of continuous quality control of the soundings plotted by the surveyors.

The disadvantage of the traditional methods is that it is labour intensive, requiring possibly two surveyors to produce each chart and, moreover, if speed of production is essential, an automated chart plotting system is undoubtedly faster. In addition, data collection must be geared to the amount of information the human surveyor can reasonably handle, which does not fully utilise the potential of many modern echo sounders. The fully automated survey system uses a personal computer to capture all the positional and depth information in a suitable format on to a storage medium. The software running on the PC will also normally provide all the track guidance, including an online trackplot, and allow the surveyor to monitor the quality of the positioning system.

The diagram below illustrates the possible configuration of such a system.

The performance of such computer based survey systems varies widely, as does their cost. As a rule of thumb the more you pay, the more the system will do. However, many of the more expensive products were designed primarily for the offshore survey market, and so their capabilities are far beyond the needs of many port and harbour users. The main criterion for most hydrographic survey applications is speed of data acquisition. A simple specification for such a system is outlined below, as are the main factors which will affect the performance of it.

1. Compute position by least squares adjustment from one positioning system at a time.
2. Indicate fix quality and residual errors in real time.
3. Provide position updates at least once per second.
4. Be capable of logging up to 10 depths from the echo sounder per second.
5. Supply fix printouts, online track guidance, and online track plot without prejudicing data acquisition.
6. Be robust, easy to operate, and easy to maintain.

The factors which will determine whether a system will meet such a specification are set out below:

Software complexity

Ensure the software does what you want it to do, not more, which may entail some speed penalty, nor less, which could make it redundant very quickly.

Software language

In the broadest terms programs, written in higher level languages, such as Pascal or C, will be faster than those written in a language such as Basic.

96   GPS: applications and implications

Computer hardware

The type and generation of machine that the software runs on has considerable effect on cycle times. The software should also be easily transportable, allowing upgrade of the computers as faster machines come onto the market.

The volume of data which can be captured by a fully automated hydrographic survey system means that the shore processing of data and subsequent production of charts must also be automated. Some ports now combine their survey data with environmental monitoring of the seabed, prevailing tides, currents and weather patterns. The objective is to build predictive models of seabed topography under given environmental conditions. This in turn allows more efficient use to be made of limited resources.

2.1.2 Dredging operations

Many ports require regular maintenance dredging, and these ports often have in-house dredging capabilities. Indeed, for many ports, it is the requirement for increasingly accurate dredging which has given the impetus to move to more advanced positioning and survey methods. Even where the dredging, services are contracted in there is still an obvious need to monitor the changes in seabed topography so that such services can be utilised efficiently.

The first part of any dredging operation must be to plan the location and quantity of material to be removed. This requires accurate information on the current seabed topography, which is then compared to the required bottom shape, resulting in a dredging plan being calculated. Successive surveys are then compared with the pre-dredge information and required bottom profile, to evaluate the progress of the dredging operations. The important elements of such surveys are accuracy, both in position and depth measurement and an ability to produce the results quickly. Ports with a lot of dredging activity tend to employ the most advanced survey systems they can support within their budgetary constraints.

Since such dredging operations often take place relatively close to the shore, they also facilitate the deployment of highest accuracy, short-range positioning systems. Instruments such as the Krupp Atlas Elektronik Polarfix, which is a laser based range-azimuth system, will give repeatable accuracies of less than one metre over a range of up to 4 kilometres. Obviously, attaining such a high order of accuracy is pointless unless similar results can be achieved on the dredging vessel. This has led to the introduction of increasingly sophisticated software packages for the dredgers, where the position of the dredging head is of primary importance. To attain repeatable positioning of better than 1 to 2 metres requires more than just a highly accurate measuring device. Any movement of the antenna unit, or reflector, on the vessel affects the subsequent position. Therefore, the movement of the antenna unit may itself be measured and the position computation designed to compensate for it.

2.1.3 Buoy movements and monitoring

Many ports need to constantly re-align their navigable channels and associated buoyage because of the effects of siltation and strong tidal streams. With due

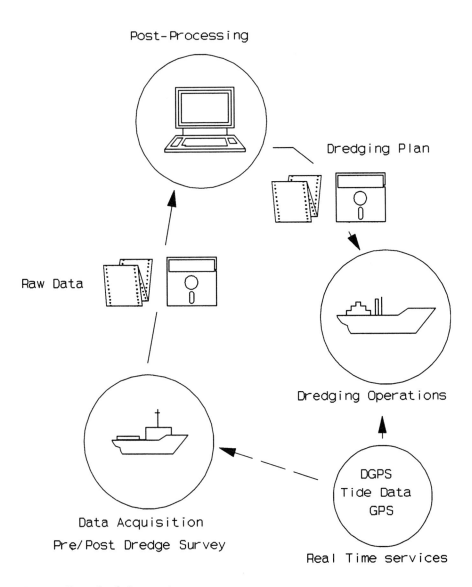

Fig. 49. Port dredging system

regard for the safety of navigation port authorities are obliged to ensure their buoyage conforms as closely as possible to that on published charts. This means they must be accurately positioned and regularly checked. There may well be additional requirements for permanent position monitoring of unmanned light vessels, critical navigation buoys and SBM Moorings. This entails a permanently installed navigation receiver on the vessel or buoy to be monitored and a data link to the shore control station.

2.1.4 Vessel navigation and pilotage

Dredging, hydrographic surveying and buoy monitoring amount to a considerable, and diverse, positioning effort for many ports. All these activities are, however, directed towards one common purpose, i.e. the safe movement of seagoing vessels into and out of the port. The port approach and subsequent harbour manoeuvring has long been recognised as a specialist task. The subsequent pilotage services are a feature of port navigation throughout the world. For many ports faced with old or suspect hydrographic data, or with rapidly changing seabed topography, it is only the skills and local knowledge of their pilots which keeps them open for business. However, such skills are not the sole prerogative of the harbour pilot as many coastal vessels and other regular traffic traditionally do their own pilotage, where permitted.

Position fixing is, and obviously always has been, an important part of the pilotage process, whether by radar, Decca or sextant for example. As the size of seagoing vessels has increased, so the navigation tolerances for restricted passage have decreased. It is now not uncommon for large vessels to undertake such passages with underkeel clearance measured in decimetres and channel width only a few metres greater than the vessels' beam. Obviously under those conditions the positioning element of pilotage assumes great importance and this is reflected in the increasing number of ports which have installed dedicated systems to aid pilotage.

There are three distinctive methods a pilot may employ to use the accurate positioning available from a port positioning service. The first method is simply to install the appropriate receiver onboard the ship and use it as one would any other navigational aid, such as radar or Decca. The pilot benefits from increased positioning accuracy and he may find other receiver functions, such as waypoint navigation, useful as well. The main drawback to this method concerns the data assimilation rate of the pilot. When piloting large vessels with small navigational tolerances, the pilot may already be at full stretch handling the information being passed to him by various sensors and members of the ships crew. Under certain circumstances, yet another navigational aid may be the straw that breaks the camel's back.

The second method is to install the receiver onboard and feed the positioning information into the ship's existing electronic chart system. A full electronic chart system will integrate compass, radar, chart, sounding data and positioning data from all sensors into a single, interactive, real-time display. The latest hydrographic data concerning the port could be transported by floppy disc and fed into the ship's system before pilotage commences. For vessels which do their own pilotage, such information could be periodically updated on each port call. Such a system will obviously be an extremely powerful tool in the hands of an experienced mariner or pilot. However, the high initial costs of such systems will mean their slow uptake by many ship operators.

Assuming, then, that the pilot requires some form of integrated display and the vessel is not fitted with an electronic chart system, his only option is to take his own aboard. This is the third method of utilizing a port's positioning service and one which is becoming increasingly popular. Such a system is unlikely to be a full electronic chart system as described above, but rather a chart display system, without the ability to overlay radar images and perhaps even without all the information one would expect on current nautical charts.

Such electronic chart display systems (ECDIS) are already available and some, such as the PINS 900 by OSL and APNAS by GEOGRAFIX, are specifically tailored to the pilotage market. A typical specification for such a system is outlined below. It is worth comparing this to the similar specification outlined for automated survey systems. In particular the factors affecting performance (e.g. software language and machine) are equally applicable to these systems.

Physical specification

1. Portable, yet rugged and shock absorbent equipment
2. Compatible with various power supplies
3. Good environmental characteristics, e.g. temperature and humidity limits
4. Good electromagnetic characteristics, it should neither be prone to interference, or a source
5. Installation time of under 60 minutes

Operating specifications

1. Position accuracy 2 to 3 metres
2. Repeatability 1 metre dynamically, 0.5 metre static
3. Ship's speed up to 4 metre per second
4. Pitch and roll up to 5 degrees on all axes

Software specification

1. Takes in positional, gyro and depth information and displays the ship's position every one second
2. Logs and plays back relevant data, to enable training and analysis of system performance

Display specification

1. Large, high resolution display; typically 19" screen with 768 × 1024 pixels
2. Operator configurable display to include Port outline, breakwaters, navigable channel and any other pertinent information
3. Up to three simultaneous views of the ship's progress at different scales, and the facility to show north up, ship up, track up or any other direction

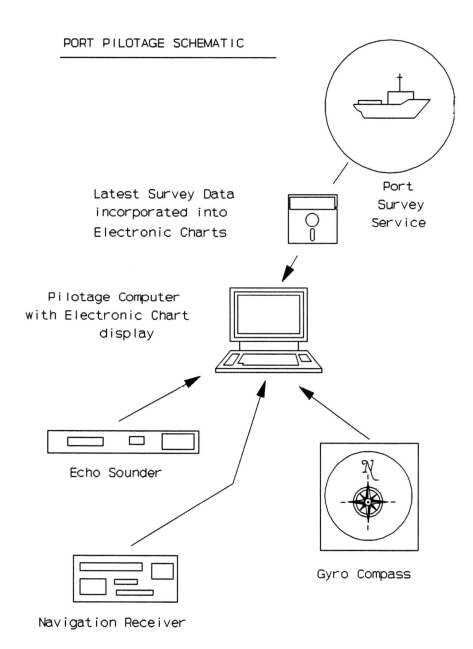

Fig. 50. Port pilotage system

GPS for port positioning 101

4. The ability to slew, zoom or rescale any of the current views
5. Alphanumeric window which can be positioned and scaled at the operator's discretion. Alphanumeric information to include gyro reading, C.M.G, speed, position quality, course to steer, distance to next waypoints, offline distances and warning messages or symbols

The specification outlined here is not exhaustive , but rather an indication of what many of the currently available systems are capable of.

2.2 The GPS port positioning service

Having discussed, in some detail, the positioning requirements for ports it is now time to look at how the GPS fits the bill. The diagram below illustrates the GPS solution to a port's positioning requirements. It shows how the one system, used in a number of different ways, can provide a total port positioning service. The question is whether the GPS solution proposed here provides an adequate service to all users, or will some specialised positioning systems remain?

2.2.1 Compatibility

For any port considering setting up a GPS service, one of the main concerns must be whether GPS is compatible with their existing systems, peripherals and positioning practices. Ideally there should be minimal disruption to current activities due to integration of GPS into current systems and retraining of personnel. The success of such an operation will depend largely on choosing the right GPS product. By the time GPS is fully operational there will be a bewildering choice of receivers and associated hardware and software products. Chapter 2 deals with this in some detail and with differential systems being detailed in Chapter 3. For the purpose of this section we will simply outline some of the main areas in which compatibility problems may occur.

Co-ordinate systems

GPS operates on the WGS84 ellipsoid. Since it is highly unlikely that a port's existing database is in WGS84 some thought must be given to converting the position data captured with the new system. Conversion software can be located in the receiver itself, in the acquisition system or as part of the database or processing system.

Communication protocols

The GPS receiver may well have to communicate both with the acquisition or navigation software and possibly to a shore based reference station within a differential network. Careful attention should, therefore, be paid to the data format output by a receiver. If it is not an accepted standard, e.g. NMEA type messages, the format should be examined to ensure that all the required information is being

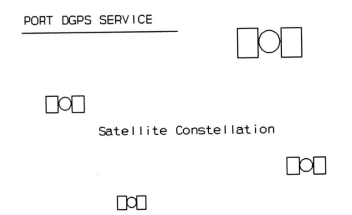

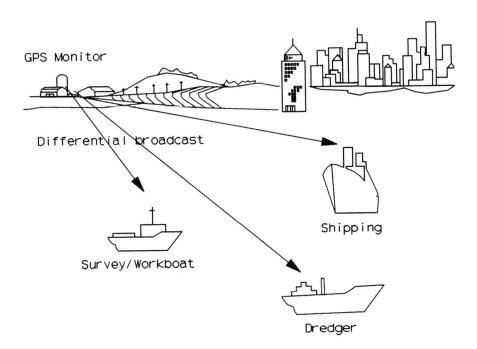

Fig. 51. Port DGPS service

transferred. This is particularly important with differential formats. Data protocol, e.g. RS232, is another important consideration. The GPS receiver must be able to physically communicate with other devices, such as computers, both quickly and with sufficient error checking to ensure data integrity.

Physical characteristics

As with any precision navigation system, considerable thought must be put into deciding the optimum antenna location for a given vessel, or monitor site. In addition, the receiver itself must be rugged enough to meet requirements, have suitable electrical characteristics, and be compatible with available power supplies.

2.2.2 Levels of service

The first part of this chapter amply illustrates both the wide variety and different levels of positioning services ports can be called upon to provide. Perhaps one of the greatest advantages of GPS is its ability to be all things to all men. By using GPS in three different ways, three distinct levels of service can be provided to users.

The three levels utilise different techniques and, to some degree, different equipment and in turn compete with different conventional systems. As each level of the GPS service is discussed in more detail below, it will also be compared to the services available from conventional positioning systems.

The precision service

The precision service would be provided by kinematic differential techniques achieving the sub-metre accuracies possibly required by dredging operation. Certainly it provides a level of accuracy never previously experienced in marine circles and even just up to a year or so ago was not considered practicable. Kinematic refers to the function of solving ambiguities in a slower dynamic environment with substantial real-time processing. The method involves using the pseudo-range, but more importantly the highly precise phase observable in the position solution under differential control. This technique is discussed more comprehensively in Chapter 5. For the purpose of this section it is sufficient to say that the accuracies attainable by this method are expected to be at the decimetre level but do require significant sophistication.

It is anticipated that this level of service will only be used for specialist survey tasks in the foreseeable future. It currently requires advanced, and expensive, processing techniques requiring major suspension of belief! However, as these techniques become better understood and the associated systems become more affordable it will undoubtedly be more widely used. Already, development systems based on low cost PCs are under trial and, although it requires the most expensive of the GPS hardware, the achievable accuracies are outstanding. In cost terms these packages cannot yet compete with microwave or laser equipment.

104   GPS: applications and implications

The accurate service

This level of service is likely to be implemented by many ports providing direct competition to existing positioning systems. It offers accuracies of between 3 and 10 metres using differential GPS techniques. The higher range of accuracies are achieved using the more expensive multi-channel GPS receivers, the lower range using the simpler and lower cost using dual channel type receivers. There are two differential techniques available, discussed fully in the chapter dedicated to differential GPS. What is important, though, is the distinction that the differential concept as applied to the port environment are often as much for the accuracies obtainable in their own right, not just as a means of building confidence. The significance of differential GPS is difficult to overstate so the following paragraphs include a review of the technique.

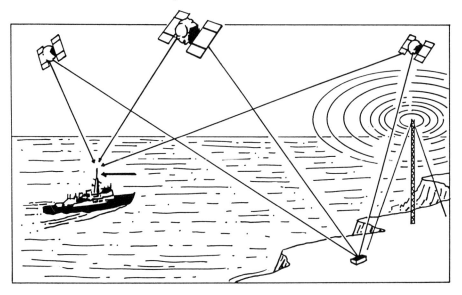

Fig. 52. A differential system

The principles of a differential service are really quite simple. A static GPS receiver, with appropriate software, is located at a known point. It monitors all the visible satellites and measures the pseudo-ranges to each one. Since the GPS transmissions include information on the precise satellite orbits, the true range to each satellite may also be calculated. By comparing the calculated and measured pseudo-ranges a correction can be determined for each satellite. These corrections are then broadcast over the differential network by a radio link and can be received by any vessel fitted with the necessary equipment. By applying the broadcast corrections to the pseudo-range measurements from his own GPS receiver the vessel improves accuracy considerably.

Although this example assumes corrections are being made to the range, they can also be calculated on the computed position given by a specific group of satellites. This may be position in terms of X,Y,Z co-ordinates or more simply latitude or longitude. Pseudo-range corrections are generally preferable, but a summary of their respective advantages/disadvantages is given below.

1. Over short ranges (0–100 kilometres between the monitor and mobile) both methods give equally good results. Over longer ranges pseudo-range differential is preferable.
2. Pseudo-range differential requires a greater standardisation of GPS receivers and differential formats.
3. XYZ differential will probably be cheaper to implement.
4. Pseudo-range differential gives each mobile the choice of which constellation to use. The XYZ method means the mobile must be using the same constellation as the monitor.
5. XYZ differential gives no alternative should a mobile be unable to track the specific selection of satellites.

Differential GPS is a service which will directly compete with many of the existing positioning systems currently sold to ports. Therefore, when deciding whether to buy a conventional or GPS positioning system for a port, it is worth considering the following points.

1 Geography

CONVENTIONAL

The size and shape of the coverage area will determine the number of shore beacons required. Awkwardly shaped areas, for example long river channels, require a large number of beacons to achieve adequate coverage. One option is to move the same set of beacons around the area to provide coverage for particular users at certain times. This, however, could hardly be considered a total positioning service.

The other obvious consideration is the maximum range required. There is a definite trade-off between range and accuracy. The higher the operating frequency of the system, the greater the accuracy and the lower the range. Microwave systems, such as Micro-Fix, are the most accurate but are line-of-site. Ultra-high frequency systems, such as Syledis, can attain accuracies of 3 to 5 metres at up to 25–30 kilometres and 5 to 10 metres at ranges up to 100 kilometres. For ranges up to 300–400 kilometres, medium frequency systems such as Hyper-fix can be used, with accuracies of at best 5 metres. The lower frequency longer range systems such as Decca seldom achieve accuracies of better than thirty metres. Beyond, the microwave systems, cost penalties are high.

GPS

To cover a given area with a differential service there is only one requirement. Namely, that all the users can receive the differential messages. This depends solely upon the transmission frequency. If, for example, line-of-site (very high frequency) transmitters are used, then the range problems are similar to that of conventional positioning systems. However, the local area can be adequately

Fig. 53. Micro-fix beacon (courtesy of Racal Marine Systems Ltd)

covered by a single transmitter. The geometry of the ground station is irrelevant with respect to a work area. This is logistically an easier option than a network of positioning beacons.

2 Operational considerations

CONVENTIONAL

Above and beyond the port's own service vessels, fewer of the port's regular users will be permanently fitted with the appropriate type of receiver. There will still be many instances where the receiver and associated antenna will need to be temporarily fitted. Since good antenna location is critical to many positioning systems, there will always be the possibility of degraded performance on such vessels.

GPS

Good antenna location is also critical to GPS performance. However, since the majority of vessels are more likely to be permanently fitted, optimum antenna location is easier to guarantee. Standardisation of formats and inter-operability of different hardware will need to be ensured.

3 Demand restrictions

CONVENTIONAL

With a few exceptions, positioning systems offering accuracies of 2–3 metres are ranging (circular) systems. These systems rely on each user being able to interrogate and receive a reply from the shore stations (transponders). By their very nature, therefore, they can only support a limited number of simultaneous users, usually no more than eight to ten. Hyperbolic systems can be used by an unlimited number of users, but suffer from geometry restrictions which are often seriously restrictive in a port environment.

GPS

The only restriction to the number of users for a GPS differential service is the number of vessels equipped with the necessary equipment to receive the differential broadcasts.

4 Reliability and availability

CONVENTIONAL

This will largely depend on the port properly maintaining both the positioning system and receivers. Adverse weather conditions, such as heavy fog, can degrade performance. Ironically this is the time when precision navigation is most essential. Conventional navigation beacons do have the advantage of being under the sole control of the user.

GPS

Only the maintainance of the receivers and differential system is in the hands of the port. Differential operation goes a long way to guarantee reliable operation, but fundamentally the on-off switch is in the hands of the American military. This concern is sometimes overstated, as with the projected phase out of Transit and Loran C there will be a responsibility to the US DOD to provide safe navigation. The nature and frequency of the system make it functionally all-weather, even in heavy fog.

The standard service

Any vessel fitted with a stand-alone C/A Code GPS receiver will be able to attain accuracies of 100 metres, 24 hours a day in all weathers. This can be considered as the standard service, for all GPS users. A port which does not wish, or cannot afford, to improve this level of service by the methods described above can still augment this standard service at little cost. By monitoring the performance of GPS at a known point and obtaining the latest information on GPS status from the appropriate sources the port can at least advise vessels in its area of the best satellite configurations to use, health of satellites and other pertinent information through normal voice channels. This can be seen as a low cost form of integrity monitoring.

3. Position and data reporting

Introduction

Position and data reporting are not new concepts. Since the advent of marine radio, ships have been able to inform head office of their position, estimated time of arrival and other pertinent information. In the opposite direction, the ship manager has been able to give commercial instructions as regards ports of call, berthing arrangements and bunkering details. The introduction of marine radio communications did not fundamentally alter the relationship between master and manager, both retaining well defined areas of responsibility and authority as regards to commercial considerations and the day-to-day running of the ship and its safety.

During the last two decades, however, this traditional relationship has been undergoing a sea change with more and more of the ship's management functions being undertaken from ashore. There have been two main reasons for this change. First, the pressure for decreased manning levels, particularly amongst the fleets of the industrialised nations, has meant an increased workload for those officers and crew remaining. It has been necessary, therefore, both in the interests of efficiency and safety, to transfer some of the decision making and general ship management functions ashore. Secondly, the introduction of secure, high capacity, global satellite communications links in the 1970s and 1980s has made such transfer of responsibility on a day-to-day basis technically feasible.

Such systems are generally referred to as electronic fleet management systems (EFMS). Their purpose is to provide real-time data communications between the fleet headquarters on shore and the ships at sea. It is worth pointing out that the concept can equally be applied to managing a fleet of trucks, railway rolling stock, or a fleet which comprises a mixture of vehicles.

The impact of GPS on electronic ship management systems will be in the fields of fleet voyage reporting, fleet performance analysis and voyage estimating, as it is these subjects which contain a spatial, or geographical, element. Such a global, continuous and reliable position fixing system offers the possibility of continuously tracking a ship over the entire navigable globe. This has obvious implications for tracking highly dangerous, or highly valuable, cargoes, and for monitoring vessels' positions in hazardous areas. A large number of different organisations may be interested in obtaining the ship's latest position including coastguard and environmental agencies, underwriters, brokers and agents.

With this in mind, the planned introduction of Standard C, and INMARSAT's proposal for a position reporting service for Standard C users, is worthy of closer investigation. This service will provide fleet managers with the means to both communicate with, and track, their fleet on a global basis. With a combined GPS and Standard C ship-board package costing less than $US 10,000 and individual position reports at approximately 10 cents this new global satellite communications system must be of interest to all shipowners and operators.

In the context of this book GPS and Standard C do appear to be natural bedfellows. They have similar system descriptions, with the divisions between ground, earth and user segments. Both provide global coverage, both are due to be fully operational at about the same time (circa 1991–1992), and they even have similar operating frequencies. Both systems also suffer from the problems of satellites being masked by obstructions, such as tall buildings in towns and cities. This means both systems have very similar operating requirements, and an environment not suitable for one will be equally unsuited for the other. Perhaps most important, both are aimed at mass markets numbered in hundreds of thousands, with probable final unit costs of only $US 1–2000. INMARSAT has been quick to spot this juxtaposition, and in early 1989 awarded a $US 250,000 contract to INMOS, the UK based semi-conductor company, to study the feasibility of incorporating a GPS unit into micro-terminals capable of operating with Standard C.

3.1 INMARSAT's Standard C satellite communications service

The new Standard C system is an attempt to bring satellite communications within the reach of all marine users. It will provide a much cheaper solution than the current Standard A system, with a smaller antenna, low weight and low power consumption suitable for fitting on all vessels from yachts to supertankers.

The current costs of a Standard C ship earth station (SES) is less than 20% of a Standard A ship earth station (SES) and it is anticipated that the relative price of a Standard C SES will continue to fall as the market expands and competition between the manufacturers increases. The initial target set by INMARSAT is for manufacturers to offer a Standard C SES, delivered and fitted for less than $US 7500. It is anticipated that Standard C will not provide the same range of communications as the more expensive Standard A system. In particular, its data rate is only 600 bit/s and it can not offer voice communication.

INMARSAT envisage three classes of shipboard equipment, class 1 being a basic Standard C communications terminal for message transfer, class 2 having the ability to switch between Standard C and enhanced group call (EGC) modes, and class 3 able to continuously monitor EGC messages while transmitting or receiving messages or data.

The enhanced group call facility

The enhanced group call (EGC) facility is available exclusively in the shore to ship direction. It will allow ships to receive messages addressed to designated geographical areas (SafetyNet) or selected groups of ships (FleetNet). SafetyNet is intended for use by national and supranational administrations concerned with the promulgation of maritime safety information. It could be used, for example, to send NAVAREA and storm warnings. The geographical extent of the broadcast area can be any size of circle or rectangle within a satellite footprint. Only ships within the designated area receive the messages.

FleetNet is intended for commercial use. The message originator, which may be a fleet manager, national authority or subscription service, broadcasts a message, which can only be received by vessels with the corresponding unique group identity code. This allows confidential and sensitive information to be disseminated quickly throughout a fleet without any fear of interception. For additional security it may be possible to encrypt the message to be broadcast.

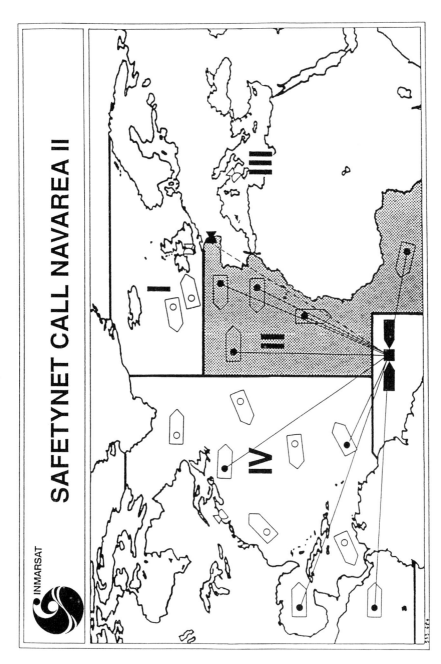

Fig. 54. SafetyNet call — Navarea II (courtesy INMARSAT)

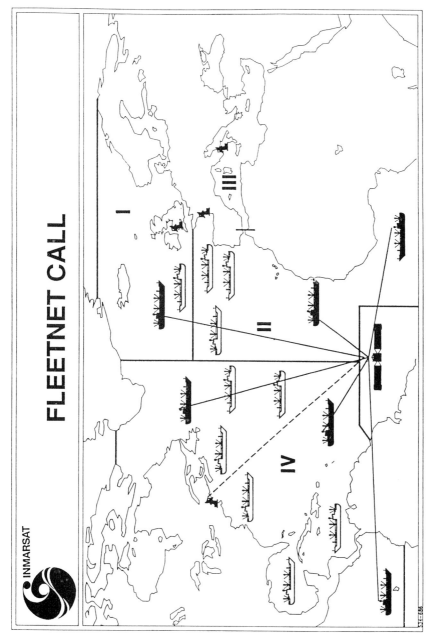

Fig. 55. FleetNet call (courtesy INMARSAT)

INMARSAT position reporting and surveillance service

This service, intended for land mobile and maritime users, will provide cost-effective transfer of position reports (with or without ancillary data) from mobiles to a base station located at, for example, a fleet headquarters. The position report messages can either be transmitted at regular intervals, transmitted at the discretion of the mobile operator, or be activated by polling from the base station. The message consisting of up to three packets of information. The first packet contains standard information, such as the position of the mobile, mobile identity, and estimated time of arrival (ETA) of the mobile at its intended destination. The two remaining packets are used at the discretion of the operator and he decides the content.

Position on the mobile can be determined by any navigation sensor (GPS, Loran C, Decca, Transit, DR etc.) which will be interfaced to the Standard C mobile using the NMEA 0183 standard. The choice of which positioning system(s) to use will depend on what additional (ancillary) data, if any, the base station wants sent back from the mobiles and how the base station manager intends to use such data. The ability of the base station manager to manipulate and analyse the data will, in turn, depend on the sophistication of the base station presentation system.

The communication link between the coast earth station (CES) and base station may consist of a packet switched data network (PSDN), circuit switched data network (CSDN), public switched telephone network (PTSN) for voice band data, and telex. Initial systems may be able to communicate with the CES using most commercially available modems, however more specific data protocols may be introduced at a later date. The capabilities of the base station presentation system may well determine which positioning system(s) is to be used on the mobile. This point is further illustrated in the three scenarios outlined below.

3.1.1 Scenario one: A basic position reporting system.

Probably consisting of a Class 1 Standard C receiver (see above), the basic system may only be fitted as a backup to the main SOLAS installation, remembering that future carriage rules may also require ships to be fitted with an electronic positioning fixing system (EPFS). Since the position reporting facility may only be used at infrequent intervals, perhaps once a day, and only utilize the basic position format, the type of EPFS is irrelevant. Indeed, position could be entered manually. The base station presentation system could be something as simple as a telex printout when the mobiles send their position reports.

3.1.2 Scenario two: A position reporting and display system

In this case the shipping manager wants more than just a position report back from the ships in his fleet. However, he may still not want the information more than once or twice a day. The ancillary information contained in the position report could take the form of an extract of the daily bridge log, including distance run, fuel consumption, wind direction and sea state. The positioning element becomes more important in this scenario. An EPFS input is desirable, and for truly global coverage GPS becomes the obvious choice.

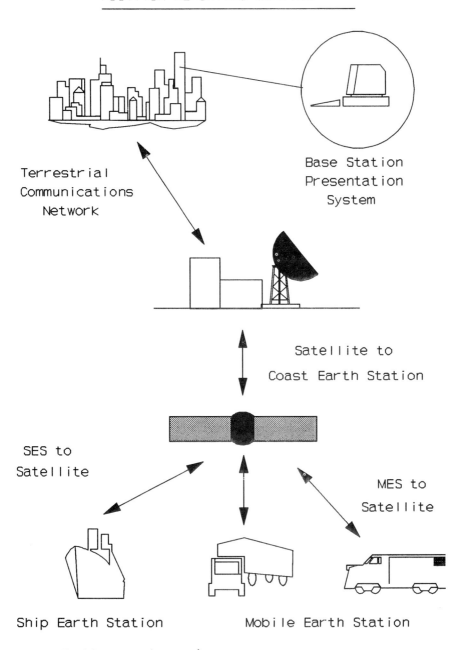

Fig. 56. Position reporting service

The base station presentation software becomes more sophisticated with the intermediate system. The type of system being proposed consists of an IBM PC linked to the CES over conventional phone lines and or telex. The presentation software on the PC should have the ability to display the position of at least 10 mobiles on a map on the screen. The positions would be displayed in real-time and the software able to log both the position and ancillary data to disk. This presentation software would also be capable of handling all communications with the CES, entailing user configurable data protocols, connecting to electronic mailboxes, and the ability to build, send, and receive standard text messages.

A number of such packages already exist, these include the CONDOR system by Geografix, which will serve as an example. The software runs on any IBM AT or 386 compatible computer, such as are found in most offices. The only additional equipment required is a modem to communicate with the CES. The system allows the use of any Hayes compatible modem, which is among the most widely used types.

The software functions provided can be divided into four main groups;

(1) Communications

This includes the commands required to communicate with the CES, initiate a fleet poll and empty the company's electronic mailbox. The mailbox contains the latest position reports from the fleet.

(2) Position display and logging

The latest positions of the fleet mobiles are displayed on a high resolution graphics chart, and an historical record for each mobile is stored on to disc. The position display commands include zoom-in and zoom-out which rescales the display centred on a given mobile or location, and history which displays the historical track of a given mobile for a given voyage. The charts provided are at a variety of scales.

(3) Position report database

When position reports are received they are stored in a relational database, which is really the heart of the system. A database is considered relational when the data stored within it is presented to the operator exclusively as tables. The database is configured to accept the optional data packets containing information such as fuel consumption, and present it in an format which is familiar to shipping managers.

(4) Reports and data export

This system does provide some reporting facilities, including printouts of the chart displays and position reports. For reports including business graphics such as pie-charts, graphs and bar-charts, the data can be exported from the database into ASCII files, into common business packages such as Lotus 123., or into an existing corporate database.

3.1.3 Scenario three: An electronic fleet management system (EFMS)

Such a system requires each mobile to be equipped with a computer and communications package. The onboard computer can be anything from a cheap and cheerful data logging device, to a full-blown ship control system. The communications are a two way process, the mobile sending back reports to HQ and the fleet manager sending instructions to the mobiles.

The September 1988 edition of Lloyd's Ship Manager reported on a number of EFMS being designed and implemented by private shipping companies. These included systems by Texaco Marine Services of Texas, Sitmar Cruises of Los Angeles and American Overseas Marine Corporation of Massachusetts. The report highlighted not only the diverse methods of passing information from ship to shore (and back), but also how much the material content of information varied from company to company. The functions of an EFMS can be extended to cover all aspects of fleet management, including fleet performance monitoring, spare parts management, fleet preventative maintainance programs, fleet payrolls and fleet personnel management. Much of this, however, is beyond the scope of this book.

With a fully automated ship, data transfer between ship and shore would be frequent. Indeed it is obvious that the amount of data transfer is directly proportional to the level of automation on the ship. The more the automation, the more information is required onshore to analyse performance, efficiency and safety. At this stage it is unlikely that the Standard C position reporting service could transfer all the data required, and so the two-way data messaging service or Standard A could be used instead, although at greater cost. The frequency of data transfer will also determine the frequency at which the vessel's position must be determined, and, at this point, an electronic position fixing system becomes an obvious requirement.

Since the purpose of transferring data back to shore is to allow detailed analysis, it makes sense to ensure such data is as accurate as possible. For historical analysis it is also essential that data are contiguous and relate to the same datum. For example, analysis of a vessel's fuel consumption against distance run is more meaningful where all the position information comes from a single system, rather than coming from a number of systems, e.g. radar, Decca and astronomical fixes.

The base station presentation software required for an electronic fleet management system will have to be a powerful and flexible tool. Not only must it be able to communicate with the CES and display and log the fleet's movements, it will also be required to analyse the performance data sent back from the ships.

A fleet management system as described above will very likely be part of an office network, acting as the interface between the ship information systems and the companies mainframe computing facilities.

3.2 Integrated fleet management

One of the most exciting prospects of low-cost navigation and low-cost communications on a global scale is the potential for integrated fleet management as illustrated above. Fleet managers can track all their mobile units, ships, trucks, barges and rolling stock, using a common system. This implies common, and hence

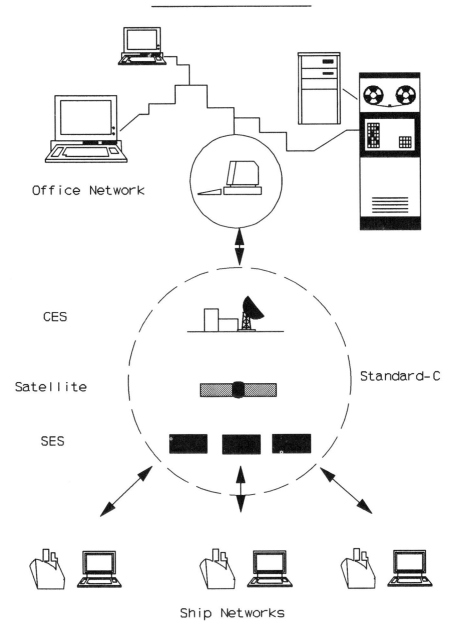

Fig. 57. The fleet network

interchangeable, hardware and software on the mobiles, which in turn allows more efficient maintainance of the system as a whole. Integrated fleet management allows more efficient use to be made of the companies' transport resources. The benefits are obvious, for example, for a company trying to arrange the sailing schedule for a vessel expecting unit cargoes from a large number of haulage units. A low-cost position reporting system is equally useful for tracking high-value or dangerous cargoes or re-routing land mobiles to avoid congested areas.

CHAPTER 5

The GPS detail

Introduction

In this section of the book we hope to provide the detail to allow a more in-depth appreciation of the technology behind the GPS systems. The explanations will be concentrated around the Navstar GPS system, although reference will be made to Glonass where it deviates from a common design.

As it will be necessary to provide substantial technical detail, the format will be in the form of short dissertations on specific topics, ordered to give a structured rendition of the subject. A more general synopsis of how GPS operates is included in Chapter 1 Navstar GPS—A System Description. This section is meant to provide full information on the system as a source of reference should it be required. In practise a GPS receiver is very simple to operate, but operation does not necessarily prescribe understanding.

1. The system design and implementation

Introduction

The global positioning systems, Navstar GPS and Glonass, have been devised to provide high accuracy three dimensional positioning, velocity information and accurate time transfer to suitably equipped users on a global basis. This is obviously fundamental to the system design.

Accuracies at the metric level can only really be provided by microwave or high frequency radio transmissions, assuming of course that electro-magnetic radio waves are to be utilised. These frequencies, due to their propagation characteristics, are generally line-of-sight or at least limited to operable ranges of a few hundred kilometres. These restrictions on their ground based operation are due to the curvature of the earth and other propagation features giving only limited coverage, near to shore. However, line-of-sight does extend for very long distances upwards from the horizon and, effectively, to infinity in the far reaches of space. To achieve high levels of positioning accuracy with these high frequency transmitters, the system designer needs to move into space, the final frontier.

The GPS systems are composed of three integral design parts—the space segment, the ground control segment and the user segment. Or in other words the satellite, the military operators and the navigator. Although proper operation of

120   *The GPS detail*

the system requires all three to be operating successfully, to a greater or lesser degree, they can be considered quite successfully as independent parts tied together by accurate time. This is the essence of the whole system.

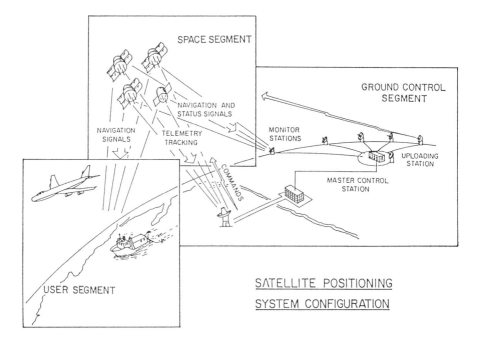

Fig. 58. Satellite positioning system configuration

1.1 The space segment

Both Navstar and Glonass have similar overall system design, almost too similar for coincidence. Currently both systems are configured to operate 18 production satellites in the finished constellation with three active spares. In the case of the American program this was reduced from an initial design specification of 21 production satellites and three active spares. This reduction occurred in December 1979 to, initially, a basic 18 satellite constellation due to a $US 500 million budget cut. In 1982, however, the three active spares were added to improve the confidence factor in the finished system. Again more recently in March 1989 approval was given to return to a twenty-four satellite constellation as soon as practicable. However, at least for the interim, the design for the initial implementation is based officially around the 18+3 design. This sequence of events is a useful reminder of the budgetary problems of such a major multi billion-dollar investment and the subsequent obvious political implications.

Orbit design

NAVSTAR GPS

The Navstar satellites (or space vehicles) are to be operated in six orbital planes in very high orbits, approximately 20,200 kilometres above the earth's surface. Three satellites are to be located in each plane in the 18 + 3 configuration, with an active spare in alternate planes. In the 21 + 3 constellation of Navstar GPS four satellites will be located in each plane. These orbit planes are circular and currently inclined at sixty-three degrees to the equator, but with one hundred and twenty degrees phasing between each plane. This is more easily understood by the example given in Figure 59. The orbital period of the Navstar satellites is just under seven hundred and eighteen minutes, resulting in the satellites passing over the same ground point each day, excepting the fact they are four minutes earlier (give or take 1.7 seconds). In the final constellation the orbit planes will be at a fifty-five degree inclination. This orbit design was developed to guarantee that at least four satellites are always in view at every point on the earth's surface twenty-four hours a day. In many instances, however, as many as twelve or thirteen satellites will be visible to a ground based user. The circular orbit design and high elevation make the system very stable in the long term with orbit variations that are relatively easy to model, in comparison, say, to low orbiting satellites.

Certain alternative orbit arrangements have also been discussed over the last few years in an attempt to produce an optimised orbit arrangement. This was to

Fig. 59. Satellite constellation Bird-cage

122   *The GPS detail*

limit the small periods of bad geometry that may be experienced by some users in some parts of the world. The only major recommendation to gain favour from these studies has been the possibility of increasing the orbit altitude by an additional 50 kilometres. This would effectively increase all-in view coverage and thus help to improve geometry.

GLONASS

The Glonass orbit arrangement is one of the more similar features of the two systems. The orbit planes are at a slightly lower altitude than Navstar at 19,100 kilometres but with a similar inclination of just under sixty five degrees to the equator and an in-plane separation of one hundred and twenty degrees. The satellites themselves are in locations in the orbit plane phased at forty five degree intervals. Glonass is also expected eventually to achieve a twenty four satellite constellation, but arranged in only three orbit planes.

The Glonass satellites currently have an orbital period of just under 676 minutes, but will only cross the same path over the earth's surface (ground track repeat) in just under eight days; thirty three minutes under to be precise. Unlike Navstar individual Glonass satellites, therefore, do not appear in the same point in space daily (minus four minutes). However, as the satellites are forty five degrees out of phase another satellite in the same plane will, making geometry a repeatable and predictable element as in Navstar.

The Satellites

NAVSTAR GPS

The satellites used in the Navstar program are especially large, as can be seen in the corresponding schematic. They are multi-purpose platforms utilised for a series of other military projects above the GPS requirement, such as atomic flash detection and location. Most of the current testing of the Navstar system has been undertaken on Block 1 development satellites. The first of these was launched in February 1978 with a total launch count of ten, six of which are still operating today with some success. Table 6 summarises the status of the Block 1 satellites up to the launch of the new Block 2 satellites.

Table 6. Block 1 Status (July 1989)

SVN	Launch date	Current status
01	22/2/78	Atomic clocks failed 25/1/80
02	13/5/78	Atomic clocks failed 30/8/80
03	07/10/78	Still operating. batteries low
04	11/12/78	Atomic clock failed, unhealthy
05	09/2/80	Reaction wheel failure 11/5/84
06	26/4/80	Operational, attitude problem
07	Launch failure	
08	14/7/83	Fully operational
09	13/6/84	Fully operational
10	8/9/84	Fully operational
11	8/10/85	Fully operational

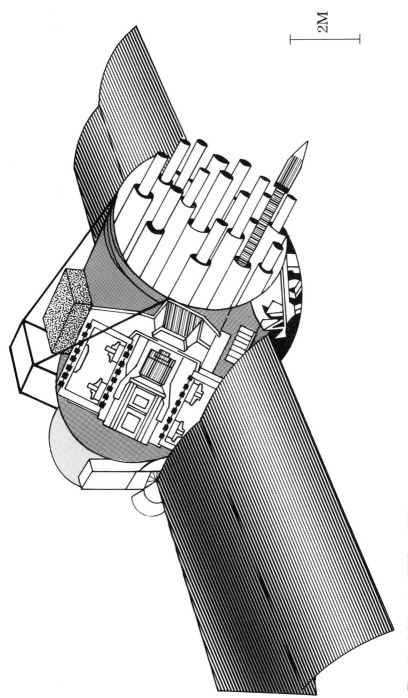

Fig. 60. A Navstar GPS satellite

124   *The GPS detail*

The reliability of the Block 1 satellites has actually been very good even if three are out of service. The design life of these test vehicles was only five years, which on average has already been achieved. In fact the oldest operating satellite (SVN 1) has actually passed its tenth birthday. Currently (July 1989) six of the Block 1 satellites are still operable although two of them may not see the year out (SVN03, SVN06). The seventh (SVN04) is operating on a quartz clock only, not considered stable enough for reliable and accurate navigation and, so, is considered unhealthy.

It is important to realise that, confusingly, there are two numbering schemes for the Navstar satellites. The first scheme is based on the launch sequence and are

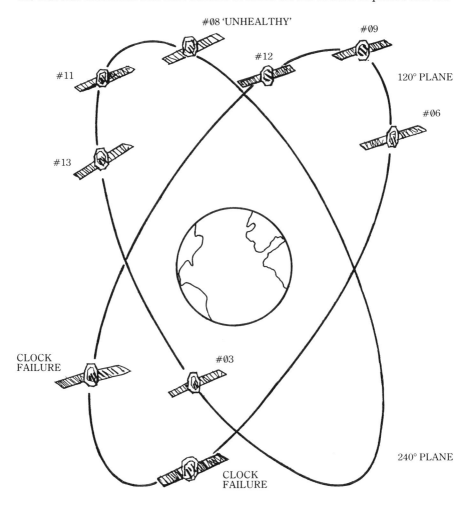

\# PRN (SVID) NUMBERS

Fig. 61. Navstar GPS Block 1 in-plane constellation

termed Navstar numbers or space vehicle numbers (SVN). This is what has been used above following the American Joint Program Office convention. However, the second and user orientated scheme, is based on the orbit arrangement of the on-line transmitting satellites. These are known by their pseudo-random number (PRN) or alternatively SV ID (space vehicle identity). These are the numbers displayed by a receiver and the scheme adopted by this book subsequently.

Table 7. Navstar numbering schemes

SVN/NAVSTAR	PRN/SVID
03	06
04	08
06	09
08	11
09	13
10	12
11	03

In February 1989 the first of the Block 2 or production satellites was launched (PRN 14). This was also the first attempted launch since the shuttle disaster three years earlier. The launch was conducted on a solid rocket booster of the Delta 2 design. The launch was quite a success, especially as it involved not only the new type of satellite (Block 2) but a new launch vehicle, a new command and control system and a new ground control facility at Colorado Springs all within a new administrative structure. The launch numbering terminology also changed from the Block 1 Navstar 1–11 sequence to the Block 2 GPS 12–40 numbering sequence. During production of this book four more Block 2 satellites were launched, confusingly called SVID 02 (Navstar (SVN)13) and SVID 16 (Navstar (SVN)14), SVID 19 (Navstar (SVN)15) and most recently in December 1989 SVID 17 (Navstar (SVN)16).

In the Block 2 family of satellites twenty eight production satellites are planned, currently under assembly at the first ever satellite production assembly line at Seal Beach, California. In addition, plans are already being drawn up for a further twenty replenishment satellites to be known as Block 2R. These will replace the Block 2 satellites as necessary and introduce some new design features. Block 2 satellites are somewhat different from the Block 1 satellites in design due to certain improvements in technology and design upgrades. However, the overall design is still, generally, similar. Electrical power is supplied by large solar panels, which help stabilise the satellite along with momentum reaction wheels controlled by powerful magnets. Battery back up is also provided for when the satellites move into earth eclipse. In fact, it is the failure of the battery back-up that is causing some problems with PRN 06. PRN 09 also has technical difficulties related to problems with the reaction wheels in maintaining a consistent attitude.

The Block 1 and 2 satellites also have only a limited supply of propellant to allow any in or between orbit manœuvres. As these can only be undertaken with limited fuel expenditure, orbit changes tend to be long affairs lasting weeks or months, during which the performance of the satellites often degrades.

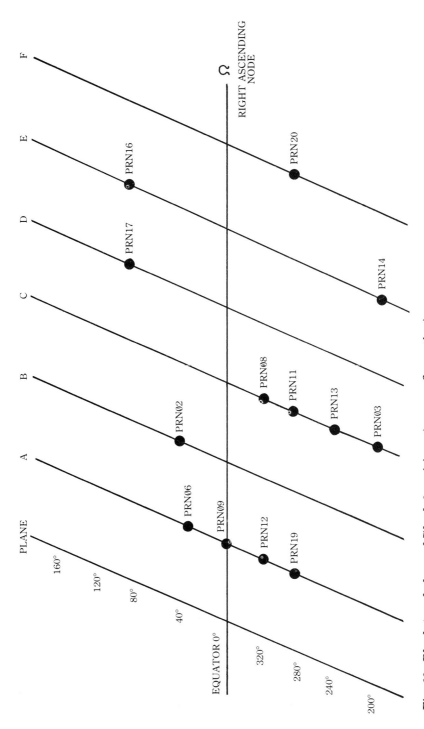

Fig. 62. Block 1 and planned Block 2 positions (test configuration)

126

GLONASS

Much less information is available on the Glonass satellites, especially with regard to the launch scheduling, although weekly new levels of information are becoming available from the USSR. Glonass has been actively operational since 1982. The first launch consisting of three Cosmos series satellites was made on a Proton launch vehicle in October of that year. Subsequent multiple payload launches have also been successfully achieved, a feat unmatched by the Americans. In fact, feasibly, the Russian system could be completely installed within a two year period, adopting a ninety four day launch window (after P. Daly, Univ. Leeds).

The Russians unfortunately appear to have had less success in maintaining the stability of their test constellation than the Americans. At the time of writing there appear to be nine healthy and usable Glonass satellites from a total launch count of twenty seven. However, it has been advised that four of these are being used purely as research and data collection vehicles, not for navigation purposes. In addition, others may have only temporarily been switched off.

Table 8. Active Glonass satellites (July 1989)

Glonass Number	International Sat Id	L1 Frequency MHz
34	1988–43A	1608.7500
35	1988–43B	1614.9375
36	1988–43C	1615.5000
37	1988–85A	1612.1250
38	1988–85B	1605.9375
40	1989–1A	1607.0625
41	1989–1B	1605.3750
42	1989–39A	1611.0000
43	1989–39B	1611.5625

Glonass satellites do not appear to rely wholly on solar or battery power, but contain an alternative power source. This gives them significant manœuvring capabilities. Orbit changes can take only a few days as opposed to months on the Navstar system. The satellites also appear to have the ability to communicate, and even possibly range, between each other, a feature only planned by the Americans for introduction into the Block 2R satellites. This has significant advantages in survivability and operability should communications to the control segment fail.

1.2 The ground/control segment

The ground or control segment refers to the ground based element of a GPS system which manages the performance of the satellites through orbital tracking, clock monitoring and therefore fundamentally is responsible for the daily control of the system. In the Navstar GPS system the overall policy-making body is the Joint Program Office. This body represents both the American army, navy and airforce and NATO. Civilian involvement includes association with the Department of Transportation and the Defense and National Mapping Agencies.

The control segment of the Navstar system consists of three main types of operational facility. The master control station situated at Colorado Springs is

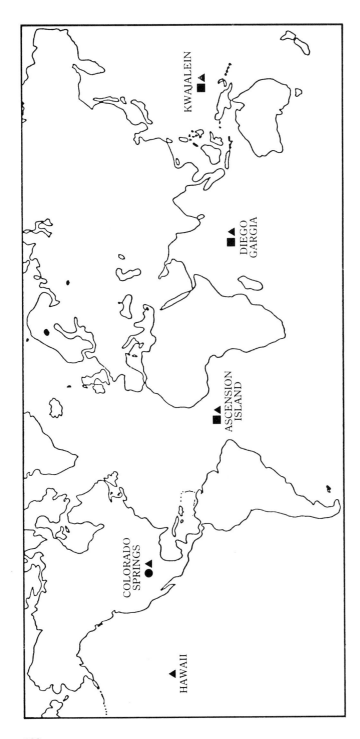

Fig. 63. *Operational ground control system Navstar GPS*

responsible for overall satellite control, navigation performance estimation and ephemeris production. Four further sites at Hawaii, Ascension Islands, Diego Garcia and Kwajalein alongside Colorado Springs are operated as monitor stations for tracking the satellites and collecting range data to produce information for ephemeris (orbit) modelling. Ascension, Diego Garcia and Kwajalein also have cosited uplink antennas to transmit navigation data and commands to the satellites. This uplink frequency is centred on 1783.74 MHz, with a downlink frequency of 2227.5 MHz. Geographically these stations produce a bracelet around the earth as can be seen in Figure 63.

Civilian GPS information centre

One of the functions of the control segment is to disseminate information regarding the performance of the satellites. This is undertaken in one sense by the setting of the health bits in the satellites to indicate their status. More information could be obtained through Navigation Notices to Users (NANUS) issued by the master control station (MCS).

In respect of the requirements for more operational and possibly real-time information a new body has been set up to disseminate information, the Civil GPS Information Centre (CGIC), a result of a US Department of Transport (DOT) initiative. Initially there was some confusion as to whom would be responsible for the organisation of such as service within the DOT, but it would now appear that the US Coastguard (USCG) have been designated the lead agency. The Federal Aviation Authority (FAA) are also involved in this service, especially with regard to the integrity issues. The compatability and inter-operability issues regarding Glonass and Navstar are within the brief of the FAA.

The functions of this information service are not particularly well defined at the moment. However, it is targeted with maintaining a data base for public access alongside a bulletin board for statements on GPS status and planned events. A future initiative may well be the setting up of an information service as part of the satellites data transmissions themselves. This is being called the Opscap (Operations Status Capability) Datapipe and will be generated for transmission both through a ground-based bulletin board and directly through L band transmissions from the satellites.

Glonass

Little information is again available about the detail behind the organisation of the Glonass system. The ground-based tracking/monitor stations appear to be located only in the Soviet Union. This will obviously limit the overall control of the system, resulting in possibly slightly less reliable health monitoring and a degraded ephemeris. For example, the satellites will be out of direct view of the USSR for up to sixteen hours. It is important to realise however that the satellites can also monitor each other which should certainly allow some integrity checking to be undertaken during these times.

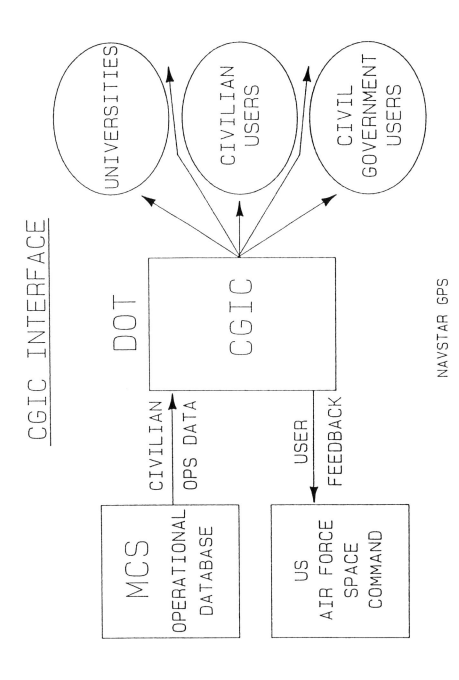

Fig. 64. Navstar GPS CGIC interface
Source: Joint Program Office, 1989

1.3 The user segment

The user segment of the GPS system includes all the elements required to successfully utilise the signals being broadcast from the satellites. In essence this means the GPS receiver (a detailed breakdown of the components of a modern GPS receiver will be discussed shortly). However, the user segment also includes such elements as the differential GPS functions and integrity monitoring already discussed in Chapter 3, Section 1.7.

The setting up of national civil users groups for disemmination of information could also be considered as part of the user segment. An example of this is the United Kingdom Civil Satellite Group (UKCSG) operated by the satellite study group of the Royal Institute of Navigation. This offers a very useful low cost, low technology information service to its subscribers or interested parties generally. Such local user groups are being set up around the world and are critical to the successful dissemination of information to the average user.

The GPS receiver design

As the GPS receiver, is in most instances, the users only involvement with GPS, this section will detail in some complexity the workings of a modern GPS receiver. A more general discussion about GPS receiver design is included in Chapter 2; dealing with the different types of receiver architecture. This section will not detail the difference between single channel and multiplexing receivers, but will provide

Fig. 65. GPS receiver in situ! (courtesy Magnavox Ltd)

132 *The GPS detail*

more depth regarding the actual hardware and types of processing undertaken inside a GPS receiver.

Major changes have occurred in receiver design in the last few years, moving from expensive analogue techniques, which dealt with the signal in its true form, to the new digital techniques, which involve the reduction of the signal to its digital (or numeric form). Much of the processing can now be undertaken in small digital microprocessors once this step is made. The example given here is only indicative of one type of receiver architecture, utilising digital techniques and an analogue-to-digital converter. It is certainly not the only design, with techniques such as hard limiting the received signal also common to receiver design.

From noise to signal

The first task for the GPS receiver designer is to get enough signal from the satellite transmissions into the receiver itself. This is usually achieved with a pre-amplifier/head amplifier in the antenna unit to boost the signal before sending it down the cable. The antennas are designed to receive all signals within the relevant band. It is the receiver that must distinguish the wheat from the chaff.

It must also be remembered that the GPS signals are very weak and indistinguishable from background noise at first and second glance. The signals are also spread over a 20 MHz band-width centred around the L1 frequency of 1575.42 MHz. This has the same effect as transmitting a much more powerful signal and also allows much more information to be incorporated into the transmissions.

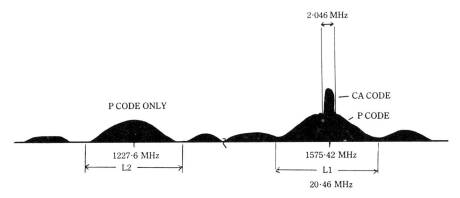

Fig. 66. GPS signal spectrum (ref. Spilker, 1980)

Transmitted and received signal strength are relatively low and for the receiver to be able to identify and extract both the frequency and, subsequently, the code transmissions, as much signal (gain) must be collected as possible at this stage. However, there is also a limit to this. Too much signal in the frequency band will overload either the pre-amp or the signal processing circuitry in the receiver. An example of this would be from Standard A or C transmissions nearby the GPS antenna. This not only means that the antenna design is critical in collecting the weak GPS transmissions, but resistance to overload from strong local transmitting sources must also be considered. This is often only properly addressed in the more expensive units.

The handling of the signal between the antenna and the receiver is also an important feature. Cabling is a major source of loss of signal gain. Pre-amplifying the signal in the antenna will get round some of these problems, but still often leaves a limit to the practical cable runs that can be laid. In response to this some receivers downconvert the signal in the antenna to a lower frequency more suitable to transmitting along a cable.

From signal to numbers

As discussed, most signal processing tasks are now undertaken in microprocessors, but to allow this the signal must be converted from analogue to digital form. Firstly, however, the signal must be made amenable to this process and this involves changing it to a frequency that can be easily managed by the new hardware.

The first task is to amplify the signal again, above that undertaken in the antenna pre-amp. In this case care must be taken not to swamp the subsequent processing with too much background noise, remembering that there is more of this than GPS signal. After this the signal is usually mixed to bring the frequency down again. to one more compatible to digital processing. Again at this point some more amplification may occur. The signal now being handled is usually between around 70 MHz to 350 MHz, dependant on the specific hardware. It has, though, been substantially changed from the initial 1575.42 MHz. These lower frequencies are selected as they are often compatible with off-the-shelf digital electronic components, but they also allow relatively simple identification of the 2MHz wide spread spectrum component of the signal, on the civil L1 frequency. In addition, some filtering may also be done at this stage to remove unwanted noise. This may be undertaken in a bandpass filter which will pass only that signal in the specific 2 MHz frequency band.

These frequencies are also really only a convenient staging post to the real processing frequency of around 1.5 to 5MHz. These are selected as they are suitable for feeding into an analogue-to-digital converter, an essential element which actually converts the signal to a digital form, i.e. in a coarse sense numbers. It is also important to select a processing frequency within the speed (frequency) of the micro-processors themselves. This generally needs to be under 12 MHz. This term is known as the clock rate and is used to define the speed of a computer. It is often quoted by the makers of desk-top computers.

From numbers to code

Effectively, the GPS receiver will now have at this point the ability to sample the substantially altered frequency of the satellites. The features of that frequency ie. the signal characteristics will have remained intact. The changes are essential to allow digital processing to be undertaken.

The primary task of the receiver is now in view—to measure ranges to the satellites. This is achieved by attempting to identify the code transmissions superimposed onto the signals.

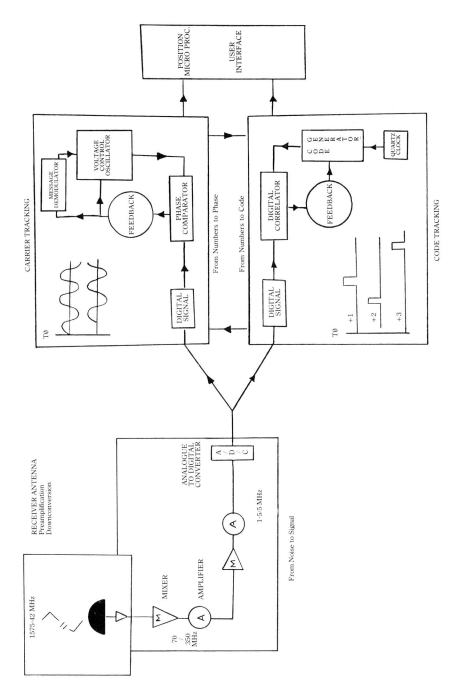

Fig. 67. Receiver design

Each satellite transmits a unique code in the Navstar system, a technique known as code division multiple access, as opposed to Glonass where each satellite has a unique frequency (frequency division multiple access). The specific technique used to modulate the code on to the transmissions will be discussed more fully in a following section.

To identify the code and thus collect all the energy from the spread spectrum signal, the receiver has to produce an exact replica of the satellite code sequence and to match the two together. To do this a technique known as auto-correlation is adopted, but this requires at least a close match to be initially achieved. The receiver must, therefore, be able to move its replica code about in a systematic way until the picture fits, a bit like trying to match the right piece in a jigsaw. Many pieces may look right, but only one fits perfectly. The function that achieves this is known as a digital correlator. Once the match is made a high correlation is seen and the maximum amount of energy is released.

Digital correlators work in various ways, but a common one is known as the early-late technique. The replica code generated closes down on to the actual satellite code transmission by straddling it with two replica codes, one too early and one too late. By moving the two estimates, advancing one and retarding the other, a correct synchronisation is achieved, which gives the same output of power in both estimator channels. This indicates correlation.

As the code sequence is moving in time, resulting from the fact that both the satellite and receiver may be moving, this correlation is a continuing exercise to some degree. This continuity in tracking is achieved through a delay lock loop, which adjusts the code replica in time to maintain a perfect match. Other digital, software functions may involve matching the expected changes in the signal to the type of dynamics the vehicle is engaged in. For example, tracking a signal on a ship will be very different to tracking it in a supersonic fighter aircraft.

From numbers to phase

To make use of this code information to produce ranges and thus position the receiver, also needs another set of information, the satellite ephemeris. This is modulated on to the carrier frequency and needs also to be extracted. Unfortunately, tracking the code does not really help to do this. This is because the navigation message is modulated on to the carrier at a much slower rate than the code. This, therefore, requires the carrier to be tracked, independently.

Tracking the carrier phase uses a similar technique to tracking the code, except that the carrier phase has a wavelength of only 19 centimetres compared to the code wavelength of just under 300 metres. To maintain lock on the phase the digital processors use a frequency lock loop, which performs a similar process to the delay lock loop in the code correlator.

As the carrier phase is modulated this does introduce a few difficulties, which requires some clever processing. This is because the modulation actually removes the energy from the signal at the exact carrier frequency. The loop which helps track the carrier is also trying to make sure that the code is tightly tracked, to help remove the confusion caused by the doppler effect on the frequency. This is a result of the differences in relative velocity between the satellite and user, causing an apparent shift in the frequency of both the carrier and the code.

From measurement to position

Once the receiver has produced the necessary information in terms of measured pseudo-range and navigation data, this is then passed to a dedicated microprocessor dealing with the position computation and, usually, the user interface as well. It is interesting to note that these functions may well take up to 75% of the receivers processing time. For more detail on these procedures please pass on to Chapter 5, Section 3, Pseudo-Ranging for Position.

1.4 The system status

The GPS systems are in a significant state of change at the moment. The Navstar system is currently experiencing a return to an active launch situation and the introduction of the production Block 2 satellites. Glonass is also exhibiting a substantial change in direction with the system being offered commercially to western users in an unprecedented move.

There is certainly no doubt about the completion of the systems, which possibly was not the case towards the middle of the Reagan administration. In fact, the Russian offer regarding the possible availability of Glonass for civil use, first officially made in early 1989, certainly expands the potential of inter-system operation. If suitable hardware becomes available over the next few years then full three dimensional global coverage from the combined systems may be achieved in advance of either one on its own.

The launch scheduling of the two systems is, therefore, of critical interest to any potential user of the two systems. This is well-published with regard to the American and obviously less so with regards to the Russian system. Table 9 lists the current launch plan for the Navstar system. One thing is obvious though. With the triple launch capability of the Proton rocket the Russians, if so inclined, could beat the Americans to full 2D and 3D implementation. The American launch scheduling still includes at least one shuttle based delivery, although the majority of the satellites will be carried by the Delta II rockets. This shuttle launch is really just to prove the fact that Block 2 satellites can be placed into orbit using this launch vehicle.

Table 9. Block 2 launch schedule

FY 89	FY 90	FY 91	FY 92	FY 93	FY 94
1 2 3 4	1 2 3 4	1 2 3 4	1 2 3 4	1 2 3 4	1 2 3 4
▲ ▲	▲ ▲ ▲ ▲ ▲	▲ ▲ ▲ ▲ ▲	▲ ▲ ▲ ▲ ▲	▲ ▲ ▲ ▲	▲ ▲
	Full 2D Capability		Full 3D Capability		

FY. : Financial year April–April
▲ : SV launches.
1234 : Quarters of financial year.

Source : Joint Program Office 01 April 1989

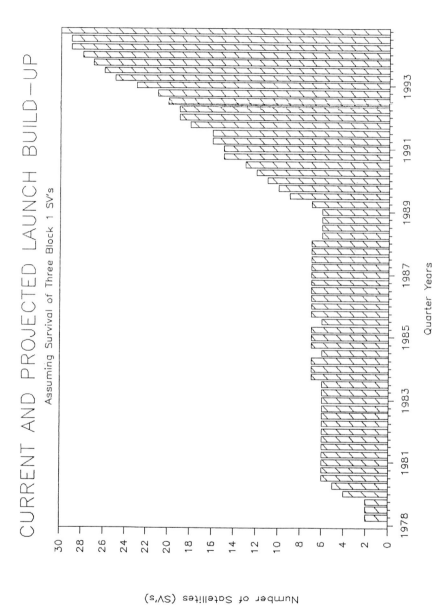

Fig. 68. Current and projected launch build-up

138 *The GPS detail*

The full 2D and 3D capabilities indicated supposedly do not include the use of any of the of the surviving Block 1 satellites. However, the ability to achieve full global 2D coverage with only a twelve satellite constellation does appear a little optimistic. Certainly it will suffer from severe geometry weaknesses. With the survival of three of the Block 1 satellites, PRN 11, 12 and 13, this target could easily be met with good world-wide geometry and possibly an earlier availability.

2. GPS: The signals

Introduction

This section will concentrate specifically on the elements of the GPS systems that allow ranges to be measured from satellite based transmissions. It will detail the signal characteristics referred to in the GPS receiver design section and throughout the book. Emphasis will be placed on the critical role that the different frequencies adopted in the GPS systems, play.

2.1 The GPS clock

All radio-navigation systems have one common feature, accurate time. In essence a stable frequency and time are the same thing in positioning, so phase comparison systems such as Decca have common roots to time comparison systems such as Loran C.

A common quartz watch is a perfect example of this relationship. The passing of a small electric current over a quartz crystal causes it to oscillate with a stable frequency, which can then be used to monitor time. By definition, a frequency will oscillate over time, producing the familiar sine wave feature. The period of this oscillation, or the wavelength, can be measured and used to clock time. In fact, the time is a meaningless concept to the watch. It is the stable, oscillating frequency which is being measured and counted. GPS uses this exact technique with even more stable oscillators than quartz. The precise transfer of time between the ground/control segment, the satellite and the users receiver is fundamental to the system operation.

To provide the most accurate frequency (time) the GPS systems use atomic clocks (also known as atomic frequency standards). These are much more accurate than a quartz clock and have drift stabilities measured in the order of a 10 E-13 per second. Each of the Block 1 satellites actually operated four atomic clocks, two known as caesium beam frequency standards and two called rubidium gas controlled frequency standards. In addition, each satellite also had very stable quartz oscillators. These four clocks not only provided back-up, but were also used to monitor one another to provide the best accuracy possible. This monitoring was used to define the clock control requirements for the Block 2 satellites.

Rubidium clocks are slightly less stable than caesium clocks in the long term, with a drift rate of 10E -12. But they also complement caesium clocks well as their short term stability, over a minute for example, is better. Under neccessity even the

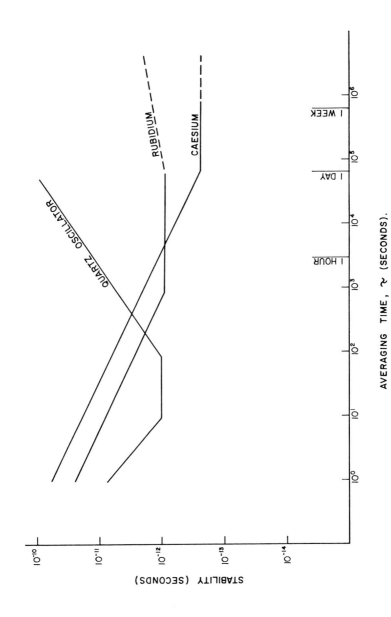

Fig. 69. Oscillator performance comparison

140 *The GPS detail*

high quality, oven controlled quartz clocks can be used to time the transmissions. This does cause a degradation in performance of the satellite, but, with frequent correction by the control segment, is still a workable option. In fact, PRN 08, a Block 1 satellite, had been operating on a quartz clock for over two years. The results determined through the trials of the relative clock performances in the Block 1 satellites gave rise to the inclusion of three caesium clocks in the payloads of the Block 2 satellites.

System time

All the clocks in the Navstar GPS system are set to operate at a frequency of 10.23 MHz. The code transmissions and the carrier frequencies are a function of this basic clock rate. Glonass again differs slightly with a basic clock rate of 10 MHz. It is essential that in each system all the clocks between the satellites are not only synchronised to this basic frequency, but drift at the same rate, thereby providing a common time reference. The clocks themselves cannot be adjusted so this must be done after the event, as a correction applied by the receiver. The monitoring of these satellite clock offsets is one of the tasks of the control segment. These corrections are uplinked to the satellites to be included in the broadcast data message (ephemeris).

If GPS was designed just to provide position then this would be the basics of the clock control in the systems. However, both Navstar and Glonass are also designed to be used for accurate time transfer. In respect of this the Navstar system time is

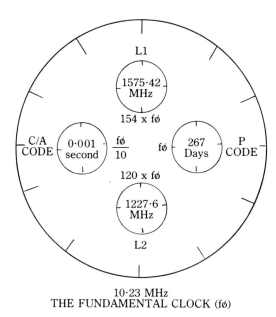

Fig. 70. The GPS clock

referenced to UTC (Universal Time Coordinated). GPS time is checked and corrected by the United States Naval Observatory to this datum on a weekly basis. UTC is currently five seconds ahead of GMT. Navstar receivers generally display UTC time.

Glonass appears to be referenced to a time datum described as Moscow time. No information, at time of publication, is available on how Moscow time is related to UTC International.

Einstein and GPS

Placing highly accurate clocks in space and orbiting them around the earth at high speeds falls foul of two of Einstein's theories. These significantly affect the GPS clock.

Firstly, as a result of the theory of special relativity, the clocks in the satellites appear to lose time, i.e. run slow as viewed from the earth. This is because they are travelling significantly faster than the control segment clocks. In GPS terms this has the effect of lowering the apparent frequency of the atomic standards. The second effect is a function of Einstein's Principle of Equivalence, part of his general relativity theory. This results in the satellite clock running faster than the control segment clocks as they are subject to lower gravitational effects, being further away from the earth's centre of mass than the control segment clocks. Unfortunately the two effects do not cancel out, with the latter effect being dominant. If left uncorrected this could cause range errors of over ten kilometres per day. To counteract this the satellite clocks are set, prior to launch, at a marginally lower frequency than the 10.23 MHz or 10 MHz standards. The calculated figures proved in practice to work, further proving Einstein's ideas.

2.2 The GPS frequencies

The relationship between the base clock frequency of 10.23 MHz and 10 MHz and the actual transmitted GPS frequencies is a critical element to the design of the systems. In reality the term "GPS frequencies" can mean a number of very different things from the clock frequency to the carrier frequencies and even the code frequencies. These are all related by design.

2.2.1 The GPS carriers

These are the most obvious of the frequencies, the actual signal transmissions emanating from the satellites. The term "carrier" is almost self-explanatory, indicating that they carry all the necessary information to produce range.

Navstar

Navstar GPS has two main carrier frequencies known as L1 and L2, used for positioning purposes. L1 is centred on 1575.42 MHz which is a multiple of 154 times

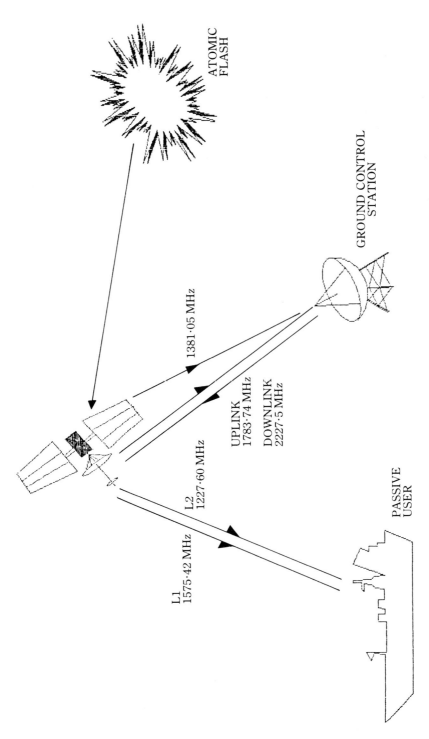

Fig. 71. *The GPS carrier frequencies*

the clock frequency of 10.23 MHz. L2 is centred on 1227.60 MHz, 120 times the clock frequency.

The satellites also operate on two other frequencies for the uplinking and downlinking of data to and from the control segment. These are 1783.74 and 2227.5 MHz respectively. One final frequency, 1381.05 MHz, is also transmitted by the satellites, but this is for their atomic flash monitoring role, having no relevance to the position-fixing service.

The term "centred" is used with respect to all these carrier frequencies, because as a result of the code and the broadcast navigation data modulations superimposed on them, the frequencies are actually spread out. This effectively changes the frequency from a high power, narrow bandwidth, to a low power wider bandwidth transmission. The bandwidth of the spread L1 signal is $+/-$ 1.023 MHz, giving approximately 2 MHz bandwidth in total. The much more heavily coded L2 signal and the equivalent component of the L1 signal has a 20.46 MHz bandwidth. The reason for these bandwidths is explained more fully in the section on the coding modulation. The full signal power is collected back from these spread, wider bands in the correlation processes undertaken in the receiver (see Section 1.3, The GPS Receiver Design).

Glonass

Glonass has a different frequency design to Navstar with regard to the carrier transmissions. In Navstar all satellites transmit on the same two carrier frequencies. The individual satellites are identified by having unique code sequences. In Glonass the individual satellites are identifiable as each actually transmits a slightly different carrier frequency. This is known as frequency division access multiple access.

The L1 frequency of Glonass is centred on 1607.0 MHz, with the L2 frequency on 1250.0 MHz. Each satellite on the L1 band transmits at an offset or channel spacing of 0.5625 MHz from his neighbour. As such SVN 01 would transmit at 1602.5625 MHz and SVN 24 at 1615.50 MHz, with the remaining 22 satellites spaced neatly between. The frequency offset/channel spacing for the satellites on the L2 band is 0.4375 MHz. The spreading of the transmissions, due to the code modulations on the L1 frequency, also appears to be $+/-$ 0.5625 MHz.

2.2.2 The codes and spread spectrum

It is rather artificial to separate the carrier frequency, the code and its frequency and spread spectrum from each other, but by doing so, and through some repetition, the close relationship between all three can be emphasised. The transmitted codes and the navigation data superimposed on to the carrier frequencies are the key to unlocking the GPS positioning concept, the means by which ranges are measured.

In both Navstar and Glonass there are two codes transmitted, one for civilian use and one for military use. Again in both cases the military and more precise code is available on both the L1 and L2 frequencies. The civilian and supposedly less precise code is available only on the L1 frequency. The way the codes are generated

144 *The GPS detail*

follows a very similar path in both cases, but as more information is available on Navstar the explanations will concentrate on this system.

It is the generation of the codes that actually spreads the transmissions of the GPS systems. Spread spectrum signals are of major importance to the military specification of the systems. They make the signals especially resistant to jamming or interference. As a result of the spread spectrum technique, when the received codes are correlated against the replica code in the receiver only the correct one will give the peak correlation and release its energy. All others will be effectively spread out in the same statistical process.

Navstar

The codes transmitted by the Navstar satellites are known as pseudo-random codes. The civilian code, the C/A code (clear or coarse acquisition) is the one decoded by all GPS receivers. The P code or precise code is available to only military users or, under special circumstances, to some favoured civilians.

The term pseudo-random code is used to refer to a transmission that is really only a succession of 0s and 1s, in what looks like random order. The fact that they are not actually random is critical to their successful application. To make use of these code sequences the receiver must actually be able to replicate them. The pseudo-random codes are also known as gold codes referring to the fact that any two code sequences have certain statistical properties ideal for the GPS ranging process. Gold codes have very high auto-correlation functions, but very low cross-correlation functions. This means that codes from different satellites will not match very easily.

The 0s and 1s of the code are created by modulating the carrier. This means that the phase of the frequency is changed by 180 degrees. It is this change that represents the movement from a 0 to a 1 in binary terms. This is best represented in diagramatic form where the change in phase, a one hundred and eighty degree shift

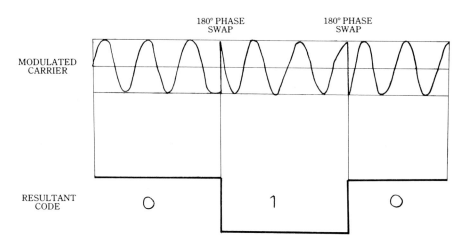

Fig. 72. Binary biphase modulation

in the wavelength form, can easily be seen. This technique is given the rather cumbersome name of binary bi-phase modulation.

Unfortunately, and very confusingly, the 0s and 1s known as code chips, created by the changes in the carrier have another incarnation known as the code state. This code state is the actual change in phase where a positive phase is known as $+1$ and a negative phase as -1. $+1$ equals a code chip of 0 and -1 a code chip of 1. This confusion is to bridge the gap between phase sign and binary arithmetic.

The ability to superimpose more than one code on to a carrier frequency comes courtesy of trigonometry. The codes can be modulated onto the carrier through a technique known as "in quadrature". This involves using the cosine element of the frequency for the P code modulation and the normal sine wave element for the C/A code modulation. Navigation data/ephemeris is modulated on to both in quadrature signals at a much lower bit rate (see section 3.2)

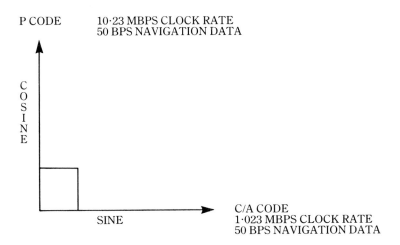

Fig. 73. GPS signal in quadrature (ref: Spilker, 1980)

2.2.3 The C/A code and the range

The C/A code (clear or coarse/acquisition) is the code used by almost all civilian GPS receivers to determine range. It also has a dual purpose in that it may also be used to help a military receiver gain access to the more accurate P code, although this can also be achieved by a knowledge of position as well. The C/A code is made up of a sequence of these 0s and 1s which have a frequency (width) of 1.023 MHz each, a tenth of the basic clock frequency. An 0 or a 1 is usually known as a bit, referring to the fact that a changeover is usually the means of passing a bit of information. However, in these code sequences they are not used to pass data but just to identify a unique progression. In respect of this they are referred to as chips not bits.

This chip frequency equates to a distance measurement of approximately 293 metres. It is, therefore, obvious that to obtain accuracies of tens of metres the receivers must be able to resolve to even closer than one bit. The code sequence

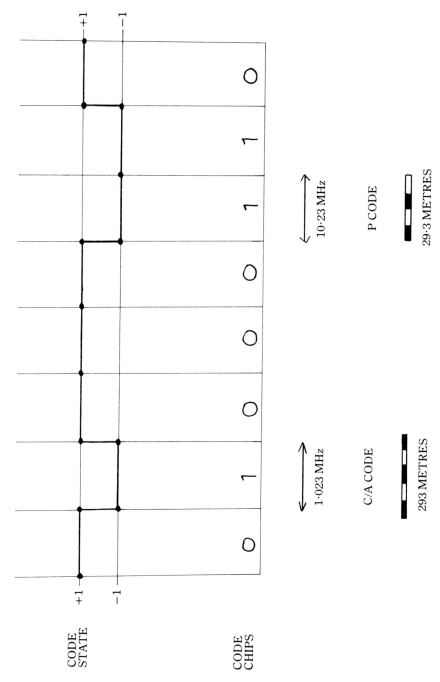

Fig. 74. The GPS codes

itself consists of 1,023 of these chips arranged in this pseudo-random order known to the receiver. This means that an entire code sequence is transmitted every one thousand of a second (a millisecond) or, in distance terms, 1,023 multiplied by 293 metres.

The receivers find it quite difficult initially to tell which code sequence they are actually measuring in, giving an ambiguity in range of about thirty kilometres. Many receivers, therefore, need to be told where they are to this accuracy, although some are now clever enough not to need even this information. This problem is exactly the same as setting up the correct whole lanes on the Decca system, which can also only measure the fractions of a whole lane.

The code chips (0s and 1s) have a frequency representative of 293 metres on the ground yet the receiver can measure the range to better than ten metres. The ability to measure to this precision has produced an interesting element to the history of the system, which is still having repercussions now.

When initially designed it was thought that the C/A code could be used to define a range measurement to only about 30 metres. This did not necessarily mean that the accuracy of the range would be thirty metres, but referred to the receivers' resolution within the chip frequency. Propagation delays would further degrade the absolute accuracy of the range. This tied in nicely with the American policy of providing a positioning service of no better than about one hundred metres. The technique adopted to measure the range was called "Code Correlation" and this was expected to allow a resolution to about one tenth (10%) of the chip frequency.

Measuring the range would involve the receiver producing a replica of the incoming satellite code and attempting to match it up. The degree that the receiver had to move its generated code to match the satellite code would equate to a difference in time between the two code makers, the satellite and the receiver. This difference in time would then be a result of two main sources. Firstly, an error in the time being measured between the satellite clock and the receiver clock. But, secondly, it would also be due to the length of time it took the satellite transmitted code to reach the receiver, i.e. range.

The accuracy that the range could be measured to the satellite would therefore be a function of how precisely the receiver could tell when it had achieved a perfect match or correlation. This is where the misjudgement was made. It was proved quite quickly that the mathematics used in correlation alongside the clean shape of the received chips gave rise to a resolution of one hundredth of a chip (1%); in distance terms, just over three metres. This allowed the C/A code to be tracked to an order of magnitude better than the Americans previously thought. This unfortunately made the difference between the C/A code and the P code less distinct, the result being the introduction of selective availability (accuracy denial) to the C/A code.

2.2.4 The P code and how

The P code (precise code) is designed as the military code and as such was constructed to give an even higher resolution. To achieve this the chip frequency is ten times higher than the C/A code at the clock frequency of 10.23 MHz. This allows resolution of range on the code to better than thirty centimetres, although propagation problems, noise and other system errors degrade it significantly below this. To make the code difficult to access the Americans gave it a repeat sequence or length of two hundred and sixty seven days, as opposed to a thousandth of a second for the C/A code. Whereas access to the C/A code is relatively straight forward and the code easy to identify as it repeats so frequently, the same cannot exactly be said for the P code.

In reality though things are not so complicated. Although the full code length is 267 days each satellite is allocated only a seven day piece of the code. A different piece is transmitted by each satellite at the same time, designed so as not to interfere with each other. Even so, it would still be very difficult to lock onto the code by chance, as effectively it still doesn't repeat weekly. At the end of the seven days each satellite is given another section of the code.

How does a receiver get access to the P code? Well, the answer is HOW, hand over word, a special instruction obtained from the navigation message, requiring knowledge of the C/A code. The navigation data can only be demodulated after the C/A code modulations have been removed. The HOW tells the receiver how far advanced into the P code sequence the satellite transmissions are, using a special timer called the Z-count. The point in time referred to by the Z-count is measured from the start of the next segment of the navigation message (a sub-frame). The Z-count is a timer which measures increments of 1.5 seconds from midnight on Saturday of each week, the changeover time for the segments of the P code. 1.5 seconds might seem a strange figure to count, but this refers to the length of time it takes for the pseudo-random code generators to generate a chip sequence.

The code generators for the P code are quite sophisticated. Two basic code generators are used to generate either a 0 or 1, the results of which are added together in binary. A further chip is generated by an intermediate code generator and the result added to the result from the basic code generators. This sequence is also being undertaken by another group of generators and the final code being constructed from addition of the two sets. This makes the resultant code very difficult to emulate and the P code sequence very difficult to understand!

2.2.5 The Y code and why?

At the beginning of this section we mentioned that only two codes were transmitted by the Navstar system and, similarly, by Glonass. So what is the Y code? Well, as the details of the P code are now well-known and documented, it does not make it particularly secure any more.

So, upon commissioning of the Block 2 constellation, the Americans intend to change the P code to a new encrypted code called the Y code. The details of this will not be available and a special code key (another microprocessor) will enable its use. These, of course, will not be freely available and some sources have it that they will be accompanied by an armed US marine guard!

2.3 Codeless GPS

A very important development in GPS technology, which has not yet really reached its full potential, is the use of the carrier phase transmissions as a means to position determination. This can actually be implemented either as a means to

Pseudo-ranging for position 149

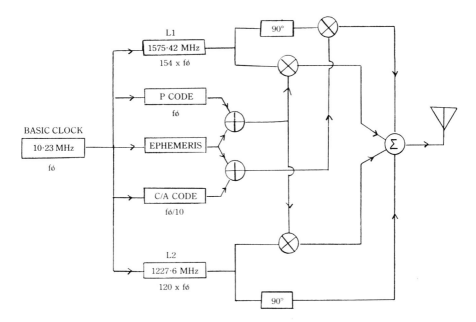

Fig. 75. The complex GPS signal structure (source: Magnavox, June 1986)

augment the code transmissions or completely independently to produce relative position solutions. As the techniques for relative positioning are necessarily quite convoluted and are unlikely to transfer readily into the marine navigation market, they will only be covered cursorily. However, phase tracking in association with pseudo-ranging does provide significant real-time improvements for the higher precision markets.

The term codeless may be a bit misleading as there are many means to achieve codeless GPS measurements. Here it is being used generally to imply the use of the carrier phase observations. One simple way of obtaining these observations is to square the incoming coded transmissions. +1 squared remains as +1 whereas −1 squared becomes +1. This results in pure carrier though the noise has been doubled as well. Knowledge of the code will still be required to take pseudo-range observations if they are needed and to decode the ephemeris.

2.3.1 Carrier aided filtering

This is more correctly called continuously integrated doppler and refers to observations on the rate of change of the carrier frequencies being used to filter out noise on the pseudo-range observations. The doppler shift is a result of the inherent movement of the satellites (see Chapter 1 on SatNav). A multi-channel receiver is capable of measuring the phase angle of the carrier to within a few degrees. As, in the terms of the L1 carrier, 360 degrees equates to only 19 centimetres, the potential accuracies are staggering.

Unfortunately, a single receiver will have no means of actually working out the number of whole carrier wavelengths to the satellite (integers). To be able to do this it would actually have to know where it is to 19 centimetres, give or take some clever mathematics. But even so the receiver will be able to measure how quickly it is moving through the waveforms, giving an indication of velocity. This technique is known as integration and can be used to indicate how quickly the vessel should be moving through the pseudo-range.

If carrier tracking is utilised in a dynamic environment there may be some problems with very fast motion changes resulting in loss of carrier lock. But to most intents and purposes it is perfectly workable on marine vessels. Continuously integrated doppler (CID) is actually used to filter the pseudo-range to remove its noise features which degrade positioning accuracy. For example, the pseudo-range, in its own right, has an accuracy of maybe only five to ten metres even after differential correction. If CID is observed this can be used to provide an accuracy of under two metres on the pseudo-range in differential mode. This can be seen rather easily in the accompanying diagram. Multi-channel or parallel receivers are really needed for this technique to be adopted.

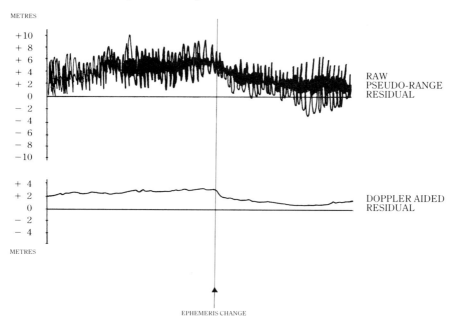

Fig. 76. The improvements of continuously integrated Doppler
Source: Magnavox June 1986

2.3.2 Phase differencing

This is the technique pioneered by the geodetic/land survey community which is capable of providing relative accuracies in the order of centimetres and, under some circumstances, millimetres. The term relative implies the point that two receivers taking simultaneous observations to the same satellites are needed in this process, similar to differential GPS. However, these receivers are primarily taking phase angle measurements and must be high precision multi-channel units. The trend is also now to use dual frequency units.

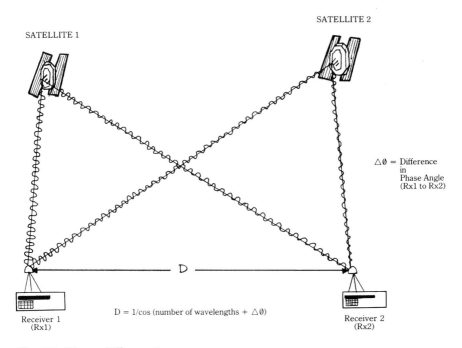

Fig. 77. Phase differencing

If two receivers take measurements to four satellites or more, differencing the measurements between different pairs of receivers and satellites can lead through a maze of mathematical processing to a difference in position between the two units. The trick here is the differencing of the measurements where all sorts of clock errors, handily, just fall out of the equation. The process even allows the number of whole wavelengths to the satellites to be calculated.

Although this technique deserves much more explanation and has provided very high accuracies in survey campaigns, it is probably somewhat beyond the scope of this book. What is relevant, though, is that, although this technique was developed for static applications only, improvements in the mathematics used now allow for one receiver to be moving, if currently only at a low steady pace.

There is no doubt that soon these procedures will work successfully on vessels, with trials already proving the capabilities. Although mariners are not interested in centrimetric accuracies certain marine engineering projects might. For example, real time monitoring of a vessel's deformation under heavy sea conditions is perfectly feasible as are platform subsidence observations.

3. Pseudo-ranging for position

Introduction

The section, The C/A code and range, gives an insight into how the pseudo-range is measured, but is certainly not complete in its argument. To complete the sequence of measuring the range and then computing the position, there are a number of significant steps that still have to be taken. As always, there are also a significant number of errors that also have to be considered.

3.1 The pseudo-range

The pseudo-range is a measure of distance from the receiver to the satellite, usually expressed in metres. It is important to realise at this stage that distance and time are synonymous. The elapsed time for the signal to travel between the satellite and the receiver (often called radio travel time) can be converted to distance simply by multiplying that time by the speed of light. The trick is, therefore, to accurately measure that elapsed time.

The term pseudo is used because the range is contaminated. For time to be accurately measured between two sites the clocks must be synchronised. The clocks between the satellites are synchronised, so ranges measured between them would actually be true ranges, a technique actually adopted in the Glonass system. But the receiver clock is not synchronised to the satellites. This gives an error which can only be resolved mathematically, hence the term pseudo-range.

The mechanics of the measurement of the range; ie. the receiver generating a replica satellite code and attempting to match them up, gives rise to the pseudo-range measurement, the degree of misalignment between these two codes. Yet the pseudo-range cannot be converted to a true range without other sets of information, namely ranges to three other satellites and accurate knowledge of all the satellites' positions in space (and time). All of this is used to help remove the time difference between the two clocks and of course to produce position.

3.2 The satellite's position

For a pseudo-range or a true range to a satellite to be of any earthly use it is necessary to know where that satellite is at all times. The GPS satellites are moving fairly quickly. It would therefore be quite difficult for the satellite to transmit a position for where it is . . . was! So, to make things easier, it transmits data for where it will be. Rather it transmits a set of information allowing the receiver to calculate where it will be. This essential set of information is known as the ephemeris or navigation data.

The ephemeris is transmitted by each satellite as an additional coded modulation of the carrier above that of the C/A code or P code. This is possible because of the mathematical simplicity of binary. The ephemeris data also called the broadcast navigation message is transmitted at a rate of 50 bits of information per second (50 bps). These really are bits, not chips, as they do carry information.

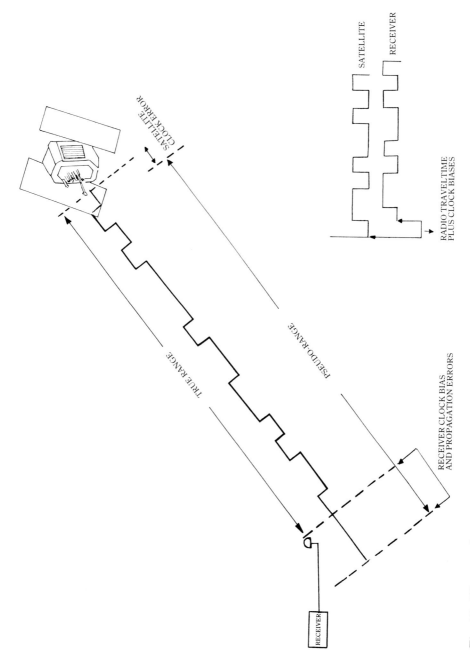

Fig. 78. The pseudo-range

154 *The GPS detail*

The bits are added by modulo 2 arithmetic to the code chip before they are modulated onto the carrier. When the modulated carrier is received it is tracked independently to extract the ephemeris from the coded sequence. The significant difference in the chipping rate allows this to be achieved. The navigation data requires a 50 Hz bandwidth for tracking, very different from the 2.046 MHz bandwidth of the C/A code. Surprisingly all these simple two state (-1 and $+1$) additions and multiplications are an extremely powerful means of passing data. Every computer is based around this arithmetic.

Kepler and GPS

Kepler was a sixteenth century physicist who described, in three laws, the movement of orbiting bodies. These laws have become the core of satellite orbital theory and are fundamental to describing a satellite's movement through space. The Navstar broadcast satellite data is correctly described as a Keplerian model. The laws are not listed here as they require more detailed explanation.

This type of Keplerian model is the most efficient and stable way of describing the change in position of a satellite, but has actually only been adopted by the Navstar program. Glonass adopts a technique based on three dimensional coordinates of the satellite being broadcast along with their differentials (i.e. their rate of change). These are much less stable over time and require frequent update. Keplerian figures do incur error over time if not updated, but to a much slower degree than the differential model.

The broadcast ephemeris is the means by which the Keplerian orbit terms are passed to the receiver. Fifteen sets of numbers (coefficients) are used to describe the orbit and these are felt to be valid for up to four hours. They are generally revised on an hourly basis by the master control segment.

Table 10. The Keplerian parameters

Time from epoch	Correction to inclination
Corrected mean motion	Corrected latitude argument
Mean anomaly	Corrected radius
Eccentric anomaly	Corrected inclination
True anomaly	Radius correction
Argument of latitude	Position in orbital plane
Argument of latitude correction	Longitude of ascending node

In addition, for a period of up to fourteen days after the event a precise ephemeris may be obtained. This is a post-tracked ephemeris indicating where the satellite actually travelled. Recently this has been upgraded in accuracy, including both pseudo-range and carrier phase observations in its assessment. The precise ephemeris is actually issued in cartesian three dimensional coordinates (x,y,z) not Keplerian terms and is available for each minute of UTC.

3.3 The computation to position

With the ability to measure pseudo-range and knowledge of the satellites position at all times, the GPS receiver now has enough information to calculate a position.

Pseudo-ranging for position 155

The solution to position is effectively that, a mathematical reduction and an exercise in three dimensional trigonometry.

The overriding problem is that the ranges are not true ranges, but are still only pseudo-ranges, contaminated by the receiver clock bias (the difference between its clock and the satellites' clocks). This is considered at this stage as an unknown. In addition, the receiver still does not know its position. This gives us three other unknowns: latitude, longitude and height. As discussed, the satellites actually work in a three dimensional coordinate system, so these are more correctly X,Y,Z.

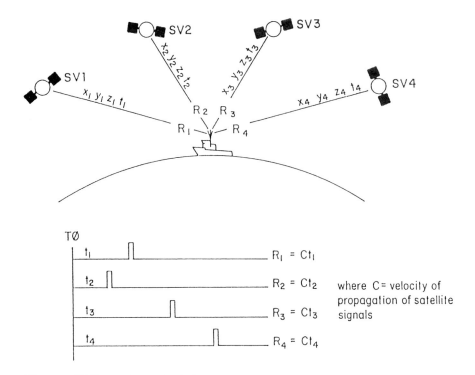

Fig. 79. GPS ranging principle

For every computation of position there are, then, four unknowns called X,Y,Z and t (for time). But by this point the receiver has also managed to acquire a whole series of knowns. Namely, pseudo-ranges to the satellites and the position of those satellites. As long as the receiver can measure as many ranges to the satellites as there are unknowns (in this instance four) then position can be calculated quite simply through as series of four simultaneous equations. This is a mathematical technique that uses a combination of known quantities to calculate a combination of unknown quantities, but it does require symmetry in its equation forms— basically the same number or more knowns to unknowns.

156 *The GPS detail*

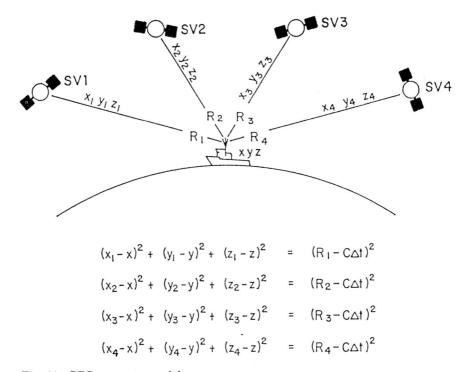

$$(x_1-x)^2 + (y_1-y)^2 + (z_1-z)^2 = (R_1-C\Delta t)^2$$

$$(x_2-x)^2 + (y_2-y)^2 + (z_2-z)^2 = (R_2-C\Delta t)^2$$

$$(x_3-x)^2 + (y_3-y)^2 + (z_3-z)^2 = (R_3-C\Delta t)^2$$

$$(x_4-x)^2 + (y_4-y)^2 + (z_4-z)^2 = (R_4-C\Delta t)^2$$

Fig. 80. GPS range to position

3.4 Position aiding

Position aiding is a technique of some significance to marine navigation and a means, during the run up to full twenty-four hour coverage, of providing extensions to the satellite working day. Aiding refers to the ability of the user to provide the receiver with additional information to ease its computational load by reducing the unknowns.

3.4.1 Height aiding

This is the most relevant of the two types of aiding to the marine navigator, and the easiest to implement. If the user's position is considered in the terms of latitude, longitude and height then the last of these is usually of little value to the navigator. In fact, the height of the antenna with respect to the sea surface is relatively easy to measure, using an ordinary steel tape.

In geodesy, the mathematics of the shape of the earth, the mean sea surface has a special gravitational significance and is referred to as the geoid. This mean surface equates very closely to mean sea level, but does not account for any large tidal variations.

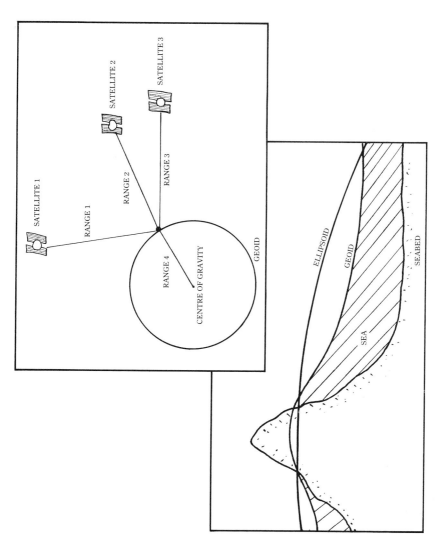

Fig. 81. Height-aiding

158   *The GPS detail*

If the receiver is given this measure of height separation from the sea surface it uses it essentially to calculate a range to the centre of the earth, which has a position in the same co-ordinate reference frame as the satellites, 0 in X, 0 in Y and 0 in Z. Effectively you have created another satellite at the centre of the earth with a known range. This reduces the number of unknowns to three and the required number of observations to three.

Operating in this mode a receiver will be able to compute position from only three observed satellite ranges. In addition, as height is now not required and is the weakest element of a three dimensional position, better navigation accuracy can be achieved and for longer. This new geoid range improves the geometry by being the only "satellite" below the user.

Height aiding is a feature that should be used by all marine navigators, but it does require that the height above the sea surface to the antenna is accurately

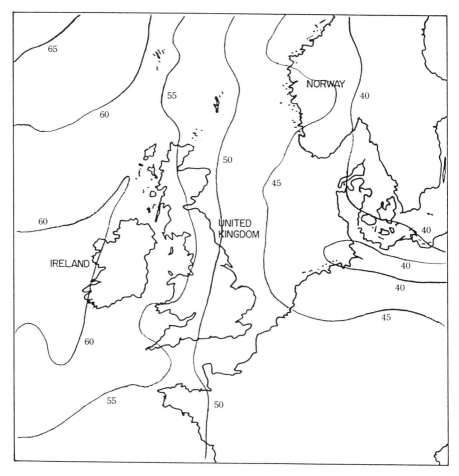

Fig. 82. WGS 84 Geoid-spheroid separation, Euroshelf (North Sea, 5 metre contour)

measured. Under some conditions a mis-measurement of height by, say, five metres will give rise to just under a five metre position error.

With some receivers an additional correction to this antenna to sea surface height (i.e. altitude) may also be required. This is because these receivers require the height to be input with respect to the spheroid not the geoid. A fuller explanation of these terms is given below in section 3.5.1, The ellipsoid. But essentially the spheroid is a much simpler (and theoretical) model of the shape of the earth than the geoid, which is complicated and created from real measurement data.

The differences between the satellite spheroid and the geoid are known, but it does vary significantly over the earth's surface. Maybe by as much as a few hundreds of metres. In some areas of the world, for example in the South China Seas, the figures change very rapidly and could cause a significant error in position if not updated frequently. In respect of this it is generally a good idea to buy a receiver that automatically corrects for these figures by holding the necessary data in permanent memory. All the operator will have to worry about then is to accurately tape the separation of his antenna above the sea surface. Other possible error sources such as tidal variations in real terms contribute little, unless very high accuracy is required.

3.4.2 Clock aiding

This is a further method of aiding the receiver that will probably be of less value to the marine community as a whole than clock aiding, although it certainly will have some applications to specialist users, such as research vessels or exploration vessels, in the run up to the completed constellation. Clock aiding requires the provision of accurate time to the receiver through the use of an atomic frequency standard, such a rubidium. This can cost in excess of $US 10,000, probably more than the receiver itself. This cost factor will obviously limit their use.

Once the offset between the satellite clock and the receiver clock has been calculated, in either a three satellite (height aided) constellation or four satellite mode, then this can be maintained by the rubidium clock down through two satellite coverage. The ability to work with two satellites, although only after a three or four satellite constellation, can substantially increase coverage hours. As more satellites are launched this facility will help achieve twenty-four hour availability well in advance of system completion.

Certain problems, however, are introduced with two satellite operations. Errors will occur over time as a function of the drift of the rubidium clock used. In addition, it is difficult to assess the quality of a two satellite fix, with most of the normal quality control factors displayed by a receiver now meaningless. Clock-aided operation should, therefore, be approached with some caution.

3.5 The position reference frame

After what now appears to be quite a laborious process the receiver will be able to compute to position. From a cold switch-on, it may take as much as twenty minutes for a receiver to reach this stage. If a receiver has been in use recently it may only take a couple of minutes. Following on from this all the tracking and computation

160 The GPS detail

procedures should be achievable, with position fixes occurring at least every two seconds. Most receivers operate at a one second fixing interval. If a receiver make is incapable of this update rate then its micro-processors must be under a heavy computational load. This suggests that the unit is already over-stretched and may be less useful in terms of forward development and upgrades. At this stage a display, or output of position to a peripheral device, should now be available. This will usually be in the geodetic co-ordinates of latitude, longitude and height. There is still one overriding concern—where does this actually place us? This question is actually more complex than at first might appear.

Normally a position is plotted on a chart either manually or electronically. Yet to be of value both position and chart must be in the same reference frame. That is, they use the same parameters for defining the shape and size of the earth. Historically, when the requirements for accuracy were less demanding, mariners took astronomic observations to determine an astronomic latitude and longitude. This was coupled necessarily with precise time measurements, the spur behind the development of precise chronometers. The procedure generally ended here with position being plotted on relatively coarse charts.

For the highest precision associated with modern point positioning, observations need to be referenced to a geodetic, earth-based reference system. This, unfortunately, necessitates an excellent knowledge of the earth's gravity field, as gravity dictates the position of the centre of the earth. This is an essential piece of knowledge for any geodetic (lat/long/height) or earth centred cartesian (x,y,z) reference frame.

3.5.1 The ellipsoid

When dealing with geodetic reference frames the single, most important question is which one? Well over two hundred years ago it was realised that the earth's shape could be best and simply described by an ellipse. This ellipse, more correctly called an ellipsoid, could be defined by two main parameters 1. Its semi-major axis (a) and 2. Its flattening (f). These are best illustrated by referring to the following diagram.

What has really caused problems is that the figures associated with the ellipsoid can and have rather arbitrarily been chosen. Scientists over the ages have all produced their own assessment of what they define as the shape and size of the earth, with, as is to be expected, little agreement. Different figures of the earth have been adopted for different parts of the globe , with even modern day satellite reference systems adopting the fashion for plurality.

Table 11. Associated figures of the earth

Ellipsoid name	Semi-major axis (metres)	Flattening (I/F)
WGS 84	6378137.00	298.2572
WGS 72	6378135.00	298.26
International	6378388.00	297.00
Clarke 1880(mod.)	6378249.145	293.465

These models of the earth also need to be anchored at some point on the solid earth's surface. This gives rise to a datum, a point where a suitable geodetic

GPS positioning quality   161

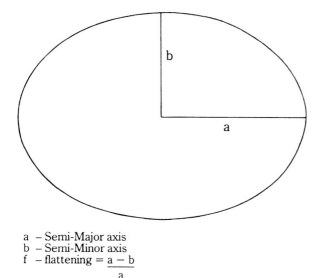

a – Semi-Major axis
b – Semi-Minor axis
f – flattening = $\dfrac{a - b}{a}$

Fig. 83. A simple ellipse

position is adopted and fixed and against which all other positions in that frame are measured. As such, a reference frame will not only have an associated figure of the earth (ellipsoid) but a datum bringing this model down to earth. Ellipsoids are more commonly called spheroids. The most common spheroids associated with navigation and maritime chart are the international spheroid of 1924 or the World Geodetic System spheroid of 1972 (WGS 72). Navstar GPS introduces a new one, the World Geodetic System of 1984 (WGS 84).

For the navigator it is obviously critical that the position output by his GPS receiver should be on the same spheroid as that marked on his navigation chart. If not, he must be able to transform his information from one to the other. This may be achieved internally in the receiver or in an external computer. To all intents and purposes GPS receivers are direct read-out devices.

3.5.2 Cartesian reference frames and spheroids

To further confuse the issue, all satellite systems actually operate in three dimensional cartesian frames, although, thankfully, this is generally invisible to the navigator. A cartesian frame is a system which has three axes, again better illustrated than explained. In most cases the origin of these axes will coincide with the centre of mass of the earth, in which situation the system will be called geocentric. They may, on the other hand, coincide with the centre of a reference ellipsoid. Co-ordinates issued for each axis define an unambiguous point in three dimensional space with no effective spatial limitations. The three axes are at right angles to another with the Z axis parallel to the mean rotation of the earth, the X

162   *The GPS detail*

axis parallel to the defined zero meridian, and the Y perpendicular to these [ref. Ashkenazi].

Cartesian reference frames make excellent sense when dealing with space vehicles, satellites or aircraft where, effectively, there is no surface of interest. This is obviously not the case with land or marine navigation, hence the use of spheroids, where navigation over a surface is the critical issue. For example, a distance expressed between cartesian co-ordinates is a straight line separation, taking no account of the fact it might be impeded by the surface of the earth, unlike a great circle which is a practical realisation of an ellipsoidal reference frame. Cartesian co-ordinates are also intuitively less easily visualised. Three dimensional cartesian co-ordinates, however, do have a further significance. They provide a suitable route to move between co-ordinates expressed in one spheroid to those expressed in another. Translations (delta X,Y,Z), rotations (on X,Y,Z) and scaling (of X,Y and Z) provide a means to move between any spheroids, irrespective of their shape and size.

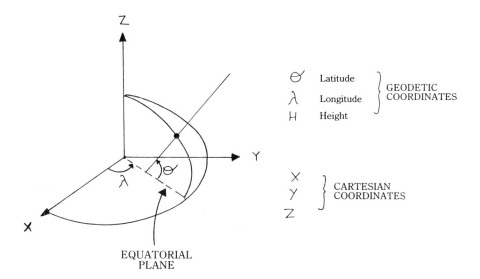

Fig. 84. Cartesian and geodetic frameworks

3.5.3 The WGS 84 spheroid

WGS 84 (World Geodetic System 1984) was first introduced to the wider survey community in January 1987 when the test and development Navstar GPS system was moved over to it from the WGS 72 reference frame. Two years later in January 1989 the established Transit (SatNav) system followed in its footsteps, now also being operated in WGS 84. Both these models have been developed to provide the best fit, globally, as possible. WGS 84 also has a very close association to the North American datum. WGS 84 was defined by the American Department of Defense and is a conventional terrestrial system (CTS). It is a modification of an existing satellite reference frame known to its friends as NSWC9Z-2. This framework was the real working spheroid and datum of the Transit system, WGS 72 was maintained, however, for wider civil use.

4. GPS positioning quality

Introduction

As important to a marine navigator as position is a knowledge and understanding of how much confidence he can place on that position. This develops intuitively from repeated use of the system and from the use of special statistical information displayed at the time of calculation of position. Both are necessary to build confidence in operation. In fact, it is probably true that currently much concern over GPS, voiced by the industry in general, is probably only due to a lack of the former. In the authors' five years' operation of the system, this confidence factor was initially of grave concern, but over the years this doubt has been assuaged. Primarily this has been a result of operating in a supervised differential mode of operation and again it cannot be overstated how important this is to providing reliable and repeatable operation. Reliability and repeatability mean safety.

Confidence itself is, surprisingly, the study of a whole branch of statistics geared to defining levels of confidence associated with certain expectations of performance and accuracy. Unsurprisingly, perhaps, it will be necessary to spend some time detailing this branch to make sense of the accuracy claims for GPS operations. In addition, the real-time factors available to quality control a position will also be detailed. This is certainly an area where improvements could be made in receiver design to make them more intelligent in their selection of satellites, as a function of real-time performance. This area will be discussed first.

4.1 Real-time quality control

Real-time quality control comes down to two main areas:
1. The ability to define how well the system is performing when everything is OK, but more importantly,
2. To define quickly when the system is not performing to specification and an error has developed

It is surprising how little information is available, in reality, to an operator of any navigation system on its performance and accuracy in real-time. Often it is just left to the operators own experience on the system's capabilities. This, unfortunately, does not protect against errors or fault conditions developing in the transmissions. This is generally left to the system controller to define at a monitoring location. This is the Ground/Control segment in the case of GPS. The system user is reliant on the vigilance and responsiveness of the control segment to protect him against such problems. This is achieved mainly through the use of the health flag in the satellite. A receiver will not work with an unhealthy satellite, unless otherwise specified. It has been noticeable to date that the control segment can be a little slow in flagging an unhealthy satellite.

164 *The GPS detail*

As has been mentioned (at length!) differential GPS allows the user to superimpose his own quality control on to the system, also providing, inherently, a capability to recover completely from many error conditions or at the least to quickly identify those against which the user is not protected. Apart from this, assessing truly the real-time performance of the system is actually quite difficult at least in the terms of defining departure from the norm. Some receivers which utilise a special type of statistical filter known as a Kalman filter, can indicate noisy positioning solutions by monitoring the filters own performance statistics. This can be a very useful tool, but does introduce some concerns of its own, especially in terms of the filter's responsiveness to rapid motion change.

4.1.1 The dilution of precision

This is one of the primary quality control indicators available to the user of a GPS receiver. Yet it only indicates the geometrical relationship of the satellites and does nothing to say whether everything is working alright. As in all radionavigation systems based on the intersection of lines of position, the relative geometry of the transmitting stations is important to the quality of the defined position. This is best illustrated by the concept of the diamond of errors shown in the accompanying diagram. The highest positioning quality and the smallest diamond of errors are provided by a combination of position lines intersecting closest to ninety degrees. The light shaded areas around each line of position are their associated possibilities of error. Nothing is ever perfect.

The dilution of precision (DOP) was the mechanism used by the GPS system planners to identify the best orbit geometry of the satellites to provide the best user geometries. GPS offers an interesting new dimension to station geometry (pun

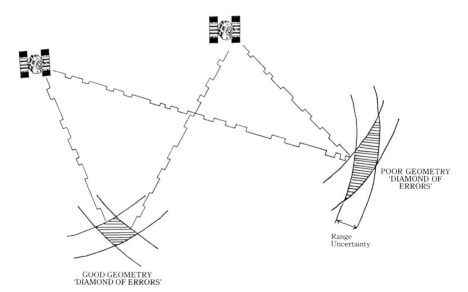

Fig. 85. Diamond of errors

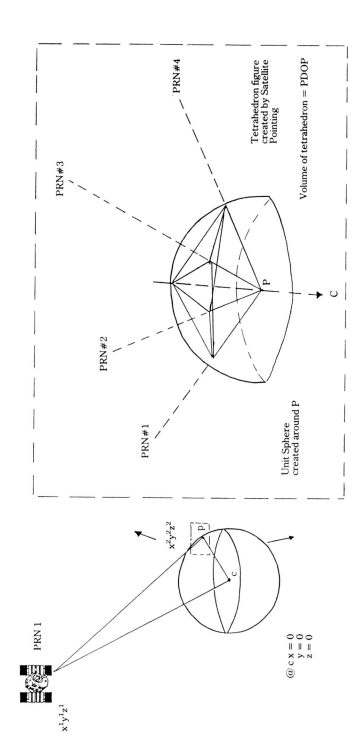

Fig. 86. The dilution of precision

166 The GPS detail

intended). Firstly, it is a system conceived in three dimensions, which certainly complicates things, and secondly, the stations are moving with respect to the user and each other with a very high velocity. This means that unlike terrestrial systems the accuracy achieved may not remain constant in a given area.

The DOP can be calculated geometrically by creating planes which relate the users position to the positions of the satellites. If these planes are placed together, a simple diamond shaped structure is created (in fact, a tetrahedron). The larger the tetrahedron, the better the geometry and the lower the DOP. Statistically the DOP is the square root of the squared errors in the position axes being considered.

The various DOPs

On its own the DOP figure is only a qualitative measure, with the prescribed numbers not being in any specific units. There is a means to relate DOP to a specific accuracy figure, but this will be returned to shortly. Firstly, it is important to realise that there is a whole plethora of DOPs, dependent on the type or state of positioning being undertaken. These are ranked dependent on the number of unknowns being calculated and are directly related to the type of position aiding adopted. The most complete DOP is the Geometric Dilution of Precision (GDOP). This is the factor used to design the orbital arrangement of the satellites. GDOP brings together the four unknowns of the system—X,Y,Z of position and t for time. This makes the assumption that the user considers time as important a variable as position. This confirms the design profile of GPS as a provider of precise time transfer as well as position.

The more frequently used DOPs are, however, PDOP and HDOP, position dilution of precision and horizontal dilution of precision. PDOP is used by those interested in three dimensional positioning (lat/long/ht.) and HDOP for just two dimensional positioning. For most purposes HDOP will be used by the marine community.

To all intents and purposes a DOP can be defined for any unknown in the GPS arena. VDOP (vertical), TDOP (time), EDOP (easting), NDOP (northing), RDOP (relative) and DDOP (differential) are all commonly occurring dilution of precision figures. Take your pick.

4.1.2 DOPs and UEREs

The UERE is yet another GPS acronym. This time it stands for user equivalent range error. In fact the UERE is the means by which the DOP can be assessed in position accuracy terms. UERE figures are also known as user range accuracies (URAs). It certainly appears that standardisation of terms is something that GPS urgently requires. The UERE figure is part of the broadcast ephemeris message of each satellite and is a performance figure, in metres, of the satellite as seen by the control segment. This is an important quality control figure, which is surprisingly absent from most receivers which feel satisfied in just providing the DOP information. A UERE figure, by definition, unfortunately cannot be real time as it is only updated each ephemeris issue. However, it does bridge the gap between the purely theoretical DOP and the reality of the actual performance of the satellite. If a satellite is working well it is usually allocated a UERE figure of between three to five metres. Figures of worse than ten metres usually indicate some problem with the control of the satellite and could alert a cautious operator to keep a close eye on the system. UERE figures appear to have been altered to keep in line with selective availability indicating a figure of 32 metres. If differential operation is adopted this need not be a problem.

If used in association with the DOP the UERE can give a feel for the positioning accuracy of the system at a particular time. This is done by a simple bit of multiplication allied to some theory of errors. If three satellites are being used to compute a position and all have UERE figures of 3 metres, they have a combined UERE figure of 5.2 metres (this is the theory of errors bit). If this figure is then multiplied against the relevant DOP (in this case a HDOP of 3.0), the resultant figure of 15.6 metres is the positioning accuracy (68% confidence level). The UERE figures and DOPs in this example can be replaced quite simply by other numbers or, if required, another satellite can be added to the sequence and PDOP used as the multiplier.

We have gone into substantial detail here as this is one of the few ways a user can determine a valuable accuracy figure for the system. But, unfortunately, the UERE figures are often not available to the operator for inspection. As these calculations are actually quite straightforward it is suprising that they are not undertaken as a matter of course in the receiver and displayed as a position uncertainty figure.

4.2 The completed constellation

As has been discussed the real-time geometry of the GPS systems is more a time based consideration, unlike existing terrestrial based systems where it is a function of geography. This is not completely the case as, even if the completed Navstar constellation, there will still be limited geographical areas where for very short periods of the day geometry might degrade beyond the accepted norm. These are generally more significant for three dimensional users, as can be seen on the figure below. It is important to stress, however, that mariners operating in height fixed mode will not experience these periods of degraded geometry.

For general planning purposes a PDOP figure of better than 6.0 has been assumed to represent the twenty-four hour positioning capability of Navstar GPS. For two dimensional positioning a HDOP figure of 3.0 is often used as the most representative geometry for the completed 18 + 3 satellite constellation. Now, if the 21 + 3 constellation seems to be confirmed then these windows of poor geometry will become even more limited.

If these DOP figures are used in association with a UERE figure of 5.2, for example, then accuracies of better than 30 metres in three dimensions and approximately 15 metres in two dimensions would appear to be possible for stand-alone navigation. These are actually not too far from the truth, assuming, of course that selective availability is not working.

These numbers also bring up a whole host of new questions. What do we actually mean by 15 metre accuracy positioning? And how representative is a general

168 *The GPS detail*

UERE figure of 5.2? Although these figures may be realistic for an undegraded constellation they still require qualification.

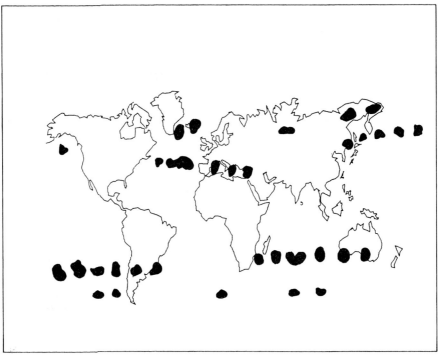

SHADED AREAS WILL HAVE GEOMETRY FIGURES WITH PDOP LESS THAN 7·0 FOR 30 MINS PER DAY
MARINE USERS WITH HEIGHT FIXED NOT AFFECTED.

Fig. 87. Geometry outages (18 + 3 active spares)

4.3 Confidence levels

An accuracy figure of 15 metres in its own right is actually meaningless, unless it is qualified. Does it, for example, mean 15 metres guaranteed ? Unfortunately this is not the case. No accuracy can be guaranteed, not even by the most persistant receiver salesman. Major assumptions are made about geometry and satellite performance, so, to indicate this, an accuracy figure should always be quoted with an associated confidence level. This confidence level indicates exactly how reliable the position is. This type of information might at first glance seem pedantic, but for operators critically concerned with safety, such as aircraft pilots, it is one of the most important subjects regarding the acceptance of a new system. It is also especially illuminating in terms of GPS where accuracy claims are often seen to be contradictory or even misleading. An incorrectly quoted accuracy could transpire to an invalid confidence being placed on a service by a user and as such compromise safety.

For confidence to be assessed the system must first be carefully studied and analysed and shown to conform to certain statistical attributes. When it has the Pandora's Box of accuracy claims can be opened. Confidence levels are directly associated to statistical figures known as standard deviations. Standard deviations attempt to define the amount of variation from an average value that a sample of data might show. These figures assume that the data exhibits a random distribution (Gaussian) which is actually not always the case in GPS terms. Even so, they are still a useful peg on to which to hang the system.

Confidence levels are derived by taking the diamond of errors one step further and considering the effect of more position lines. This gives an error ellipse figure, identical in form to the simple ellipse illustrated in fig. 83. Error ellipses are determined with reference to a circle of derived errors as shown below.

A worked example, using selective availability, will clearly illustrate these errors. Selective availability has been quoted officially at two levels—100 metres and 50 metres. These are not actually different figures, just difference confidence levels. The assumption is that these accuracy figures refer to all three dimensions. The 100 metre figure refers to a 95% confidence level, also known as two standard deviations (2o). This means that 95% of the time the positioning accuracy will be within that accuracy envelope, relative to the WGS 84 spheroid. How bad it is the other 5% of the time is not necessarily indicated. The 50 metre figure is at a 68% confidence level, or one standard deviation. This figure not only sounds better, but is often the most frequently used. Yet it is important to realise that 68% also means that just under a third of the time 50 metres will not be achieved.

Equipment manufacturers have actually gone one step further. The achievable accuracies of most GPS receivers are quoted using figures known as spherical error probabilities (SEP) or circular error probabilities (CEP). These refer to three and two dimensions respectively. Although these figures sound nicely technical they are actually only at the fifty percent probability level or 0.67 of a standard deviation. In these terms selective availability would give you a 33.5 metre positioning service.

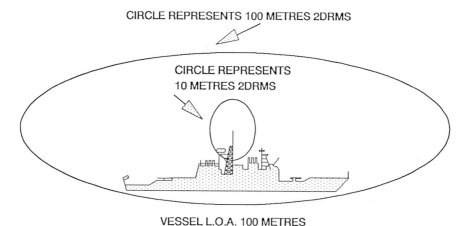

Fig. 88. The GPS accuracies

So, effectively, by just not defining the confidence level, the system accuracy has apparently improved from 100 metres down to 33.5 metres. In truth, the ninety-five percent confidence level is the most realistic and should be utilised for all planning purposes and discussions. One hundred percent confidence can never be achieved.

CHAPTER 6

Standard C detail INMARSAT and global satellite communications

1. History in brief

Telstar, the world's first commercial communications satellite, was launched on the 10th July 1962. It was a low orbit vehicle, varying in altitude between 600 and 3500 miles, and could only be used for approximately half an hour each day. Although of limited use as a communications platform, its launch did have the effect of galvanizing into action both governments and the communications industry, which led directly to the establishment of Intelsat in 1964. The International Satellite Telecommunications Organisation (Intelsat) started off with 11 members and launched its first satellite in April 1965. Intelsat 1 (or Early Bird) was a geostationary platform at an altitude of 23,000 miles, and was quickly followed by Intelsat 2 in 1966 and Intelsat 3 in 1969. Although widely used for business calls, radio and television broadcasts etc., these platforms were not intended primarily for deep sea communications and were located to cover the continental land masses.

In 1976 Comsat General, backed by the United States Navy, placed three communication satellites into geostationary orbits over the Atlantic, Pacific and Indian Oceans. These Marisat platforms were specifically intended for use by the shipping industry. However, the high initial cost of the ship earth stations (SES), in excess of $US 150,000, meant a slow uptake by civilian users. In the same year that the Marisat communication system was implemented (1976), the International Maritime Satellite Communication Organisation (INMARSAT) Convention and Operating Agreement was adopted by the International Maritime Organization (IMO). INMARSAT came into being in 1979 and began operations in 1981. INMARSAT leased the Marisat satellites from Comsat General and took over operation of what they renamed the Standard A service in 1982.

2. The INMARSAT organization

In December 1988 Czechoslovakia became the 55th member country of the INMARSAT organization. The membership includes all the major industrialized countries of both East and West including USA, USSR, Peoples' Republic of China, Japan, East and West Germany, France, Italy, and the UK. Each country has a signatory, often the national communications company, who is that country's representative in the council. The amount each signatory invests in the

172 *Standard C detail*

organisation varies considerably, with the USA, UK, Norway and Japan accounting for approximately 66% of the total. The proportion each member country pays to INMARSAT, required to establish the satellite systems and meet operating expenses, is based upon the amount of use each country makes of the INMARSAT System. The proportions paid are reviewed annually and INMARSAT undertakes to pay interest on the investments, and where possible, also repays capital expenditure.

INMARSAT derives its revenue by charging the coast earth stations for the use they make of the system. Each coast earth station/operator determines the charge to the end user. INMARSAT is, therefore, in the business of selling capacity on its satellites, rather than providing an end-to-end service.

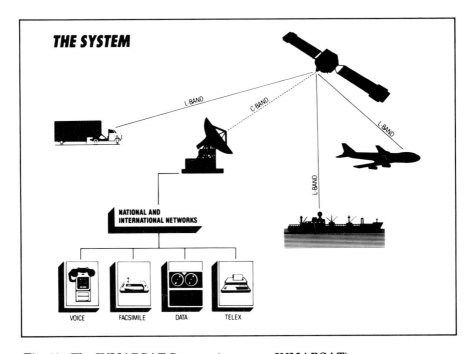

Fig. 89. The INMARSAT System (courtesy INMARSAT)

INMARSAT currently provides two main maritime services on its satellite system, the Standard A and Standard C services. Both these systems are discussed in more detail below. However, more emphasis is placed on the new Standard C system because of its potentially close relationship to GPS. The INMARSAT System (comprising Standards A and C) offers global satellite communications between 75 degrees North and 75 degrees South, where the mobile earth station is located in one of three ocean regions, as illustrated below. In addition, the proposed Standard B and Standard M services to be introduced in this decade will further enhance global communication services.

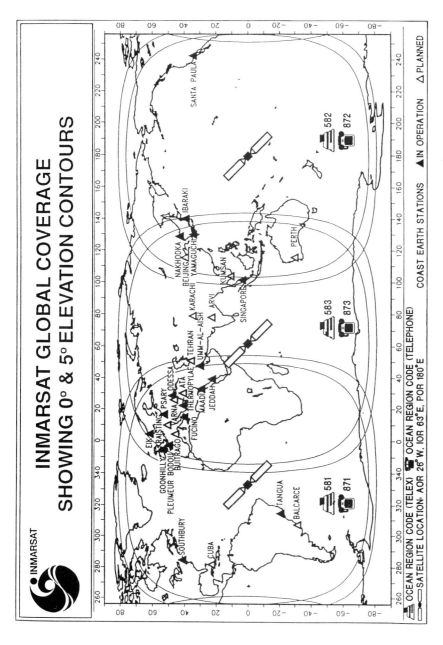

Fig. 90. INMARSAT global coverage showing 0° and 5° elevation contours (courtesy INMARSAT)

3. The INMARSAT Standard A service

The number of Standard A users has grown steadily since 1982, at an average of approximately 1000 per year, there now being some 8000 INMARSAT or Marisat approved terminals worldwide. This has been a direct reflection of the downward trend in SES installation prices, which now cost in the region of $US 35–50,000. Of the 8000 SES worldwide some 10% are for use on land and, certainly, non-maritime sales are set to increase with the introduction of aeronautical data and voice services. British Airways began such a service in February 1989 on their North Atlantic route, worldwide services being available from 1990. Approximately 60% of ship earth stations are installed on merchant ships (bulk, container, general, tanker and LGC), with the next most significant users being fishing vessels (8–9%) and Yachts (about 7%). Over half the SESs are fitted to vessels registered in the USA, Panama, Japan, Liberia and the UK (ranked in order), all of whom are INMARSAT signatories. The continued steady growth in Standard A SES installations can largely be attributed not only to this steady drop in cost, but also to their increasing use as a secure data link for office based shipboard management.

If the ship earth station is fitted with a modem and computer, a range of data carriers are available. These include a 2.4 kilobits per second (kbit/s) carrier in the telephone band, medium speed data links at 9.6 kbit/s and high-speed ship-to-shore data transfer at 56 kbit/s utilising packet-switched data, (Ocean Voice, July 1988).

A whole range of off-the-shelf ancillary services have built up around the Standard A system. Some are provided direct from the coast earth station (CES) and include options such as national and international directory enquires, technical assistance, credit card calls, store and forward messaging and telegram services. There are also six distress and safety related services like medical advice, medical assistance, meteorological reports, navigation hazards and warnings, and ship position reports.

In addition to these there are also the value added services provided by commercial companies. These often take the form of a database, requiring the ship earth station to be connected to a computer and a modem. These services include, for example, a global weather information service operated by Universal Weather and Aviation in Houston, Texas and a news distribution service provided by Oceansat TV based in the UK. The latter costs $US 900 per month for a twice daily broadcast seven days a week. Weather services are undoubtedly the most common value added services and in addition to those mentioned above, weather information is available from the US National Weather Service, Kavouras Inc. in Minnesota, WSI Corp of Bedford, Massachussets, Navtech Inc. in New York State, Sea Ice Consultants Inc. of Maryland and Metroute from the UK Meteorological Office.

Oceanroutes of Sunnyvale, California, are one of the longest established weather routing services and they make extensive use of the Standard A telex. The cost of the service to the customer depends on the type and size of vessel and the route, but is not affected by the number of calls sent or received during the voyage. The cost for an average North Atlantic crossing is estimated to be just $US 560 (*Ocean Voice*, January 1989).

3.1 The Standard A ship earth station (SES)

A Standard A SES consists of two parts, the above-deck components and the below-deck components. The above-deck components typically consist of a gyro-stabilized phased array or parabolic antenna, a solid state L-band power amplifier, an L-band low noise amplifier, diplexer and low-cost radome. The antenna is, typically, 0.85 to 1.2 metres in diameter and requires a stabilized platform to continually track the satellite during the pitch and roll of a vessel's movement. Below-deck equipment is made up of an antenna control unit, communications electronics (for transmitting, receiving, access control and signalling) and a variety of communication devices, which may include telephone, telex, modem and computer.

The SES uses two frequencies for communication with the coast earth station via satellite. Messages are transmitted at 1.6 GHz, and received at 1.5 GHz. The number of SES terminal manufacturers has steadily increased over the years. There are now some 12 commercial producers of ship terminals worldwide, of which four, Magnavox, Mobile Telesystems, Sperry, and Radar Devices are American, three, Anritsu, Japan Radio Corperation and Toshiba are Japanese, two, Marconi and STC in the UK and one apiece in Germany (Dornier), Norway (EB NERA), and China (Shijiazhuang).

Of the four top selling SES models, three, the JUE-35A, JUE-35B and JUE-45A, come from the Japan Radio Corporation (JRC), with the JUE-35B vying for top

Fig. 91. A Standard A radome installation (Betsy Ross Ade) (courtesy INMARSAT)

Fig. 92. Below-decks installation (courtesy INMARSAT)

spot with the Saturn 3S SES from EB NERA of Norway. Sales of these four models make up over 30% of the total market share.

One of the most important recent developments in Standard A ship earth station design has been the introduction of multi-channel models. Prior to their development an SES was restricted to one station identity code or call sign. This meant that although the terminal was capable of handling a variety of different communications functions, including telex, voice data and facsimile, it could only use one of these modes at a given time. Multi-channel SES models couple a number of SES terminals through the same antenna and allow simultaneous communications transfer to take place. The cruise ship *Norway* was the first vessel to be fitted with such a multi-channel system, being equipped with four Magnavox Mx-2400 SES coupled through a large 2.2 metre antenna.

3.2 The Standard A coast earth station (CES)

The table below shows existing and planned coast earth stations to operate with the Standard A service. Those marked with one asterisk (*) are planned to provide aeronautical services in addition to marine services. Those marked with two asterisks (**) are planned to provide aeronautical services only.

Coast earth stations provide the gateway between the world's terrestrial telecommunications network and the Standard A system. The coast earth stations are owned and operated by the signatory organisation for that state. This could be a national communications company, e.g. The National Telecommunications Organization of Egypt (ARENTO), a designated corporation, e.g. British Telecom for the UK, a government ministry, e.g. the Office of Maritime Economy in Poland, or a dedicated national satellite organisation, e.g. the Indonesian Satellite Corporation (PT INDOSAT).

The types of communication interface to the terrestrial network provided at each coast earth station is at the discretion of the operator. However, Telex and Voice are mandatory for Standard A coast earth stations and most provide facsimile and data transfer interfaces. Each satellite ocean region is controlled by a network co-ordination station (NCS), which manages the traffic flow within that ocean region. An example of the management task is keeping a register of all ship earth stations within the ocean region. Each coast earth station (CES) is provided with a copy of the register for each ocean region and so can determine whether to process a call originating from its terrestrial hinterland or, if it can not process a call itself, where to redirect the call so it is not lost. The network control stations for the three ocean regions are Goonhilly (UK) for the Atlantic Ocean region (AOR), Yamaguchi (Japan) for the Indian Ocean region (IOR) and Ibaraki (Japan) for the Pacific Ocean region (POR).

A typical coast earth station consists of a parabolic antenna approximately 11–14 metres in diameter. Signals are transmitted to the satellite at 6 GHz and received at 4 GHz. The L-band frequencies used for satellite to SES communications are used by the coast earth stations to transmit (1.6 GHz) and receive (1.5 GHz) network control information.

Table 12. The Standard A coast earth stations (CES)

Country	Location	Coverage region	Operational status
Denmark, Finland Norway, Sweden	Eik*	IOR	In operation
Japan	Ibraki	POR	In operation
	Yamaguchi	IOR	In operation
Singapore	Singapore*	POR	In operation
	Singapore*	IOR	N/A
UK	Goonhilly*	AOR	In operation
	Hong Kong	POR	N/A
USA	Santa Paula*	POR	In operation
	Southbury*	AOR	In operation
Kuwait	Umm-al-Aish	AOR	In operation
France	Pleumeur Bodou*	AOR	In operation
Brazil	Tangua	AOR	In operation
USSR	Odessa*	AOR	In operation
	Odessa*	IOR	In operation
	Nakhodka	IOR	In operation
	Nakhodka	POR	In operation
Italy	Fucino	AOR	In operation
Greece	Thermopylae	IOR	In operation
Saudi Arabia	Jeddah	IOR	In operation
	Jeddah	AOR	1989
Poland	Psary	AOR	In operation
	Psary	IOR	In operation
Egypt	Maadi	AOR	In operation
Turkey	Ata	IOR	1989
	Ata	AOR	1989
China	Beijing	POR	1989
	Beijing	IOR	1989
Korea, Republic	Kumsan	POR	1989
India	Aarvi	IOR	1990
Germany, FR	Raisting	AOR	1990
Argentina	Balcarce*	AOR	1990
Canada	Weir**	AOR	1990
	Lake Cowichan**	POR	1990
Australia	Perth*	IOR	1990
	Perth*	POR	1990
Spain	Buitrago	AOR	1990
Cuba	N/A	AOR	1990
Iran	Tehran	IOR	N/A
Pakistan	Karachi	IOR	N/A
UAE	N/A	IOR	N/A

AOR = Atlantic Ocean region   POR = Pacific Ocean region   IOR = Indian Ocean region

Source: *Ocean Voice*, January 1989; *Aeronautical Satellite News*, December 1988.

3.3 INMARSAT's current (first generation) space segment

INMARSAT currently operates communications capacity on eight satellites in geostationary orbits. These are the Marecs A and B satellites, three Marisat satellites and mobile communications packages on three Intelsat V satellites. Five of these satellites are maintained as operational and back-up spares, the other

Fig. 93. CES parabolic antenna, Fucino, Italy, (courtesy INMARSAT)

three being the prime operational satellites. The satellites are at an altitude of some 63,000 kilometres and the position of the three prime satellites give virtual global coverage, except for the extreme polar regions. Spacecraft control for all three generations of the INMARSAT satellites is handled at the operation control centre (OCC) located at INMARSAT's London Headquarters.

Table 13. INMARSAT's current (first generation) satellites

Ocean	Spacecraft	Location	Launch date	Status
AOR	Marecs-B2	26 W	9-11-1984	Operational
	Intelsat V-MCS B	18.5 W	19-05-1983	Spare
	Marisat-F1	15 W	19-02-1976	Spare
IOR	Intelsat V-MCS A	63 E	28-09-1982	Operational
	Marisat F-2	72.5 E	14-10-1976	Spare
POR	Intelsat V-MCS D	180 E	04-03-1984	Operational
	Marecs-A	178 E	20-12-1981	Spare
	Marisat F-3	176.5 E	09-06-1976	Spare

AOR = Atlantic Ocean region   IOR = Indian Ocean region   POR = Pacific Ocean region

180 *Standard C detail*

For the first generation of satellites, discussed below, the OCC acts as a coordinating centre for the technical control centres operated by each of the space segment suppliers, i.e. those companies which built/launched the original satellites. Both second and third generation satellites, however, will be directly controlled from the OCC. The operation control centre is also the place where ship earth stations are commissioned for operation within the INMARSAT system.

Marecs A & B2 are leased from the European Space Agency (ESA), and are production versions of the earlier MAROTS experimental spacecrafts.

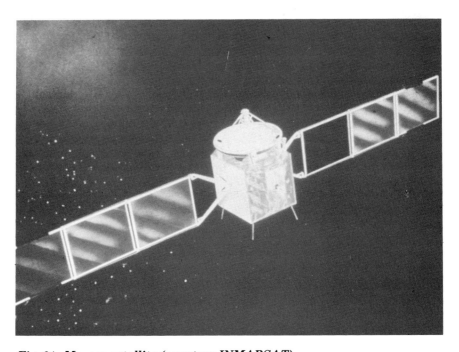

Fig. 94. Marecs satellite (courtesy INMARSAT)

Launch weight: 1,006 kg
Height: 2.56m
Solar array span: 13.8m
Type: 3-axis stabilized
Capacity: Approximately 50 2-way voice circuits
Launch vehicle: Ariane
Manufacturer: British Aerospace/Marconi

Intelsat V mobile communications sub-systems are fitted to four of these satellites and leased from the International Telecommunications Satellite Organisation (Intelsat) and leased by INMARSAT.

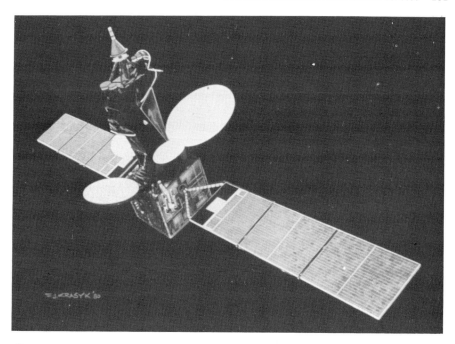

Fig. 95. Intelsat V satellite (courtesy INMARSAT)

Launch weight: 1,970 kg
Height: 6.58m
Solar array span: 15.59m
Type: 3-axis stabilized
Capacity: 30 2-way voice circuits
Launch vehicles: Atlas Centaur/Ariane
Manufacturer: Ford Aerospace and Communications

Marisat, as described earlier, these spacecraft were taken over by INMARSAT in 1982 from Comsat General, who were their operator until that date.

Frequency allocation

Ship-to-satellite communications from 1626.5 MHz to 1645.5 MHz (19 MHz bandwidth)
Earth-to-satellite (all users) distress and safety messages from 1645.5 to 1646.5 MHz
Satellite-to-ship communications from 1535 to 1544 MHz (9 MHz bandwidth), which will be increased to include 1530 to 1535 MHz from the first of January 1990
Satellite-to-earth (all users) distress and safety messages from 1544 to 1545 MHz
CES-to-satellite communications at 6000 MHz (6 GHz)
Satellite-to-CES communications at 4000 MHz (4 GHz)

182 *Standard C detail*

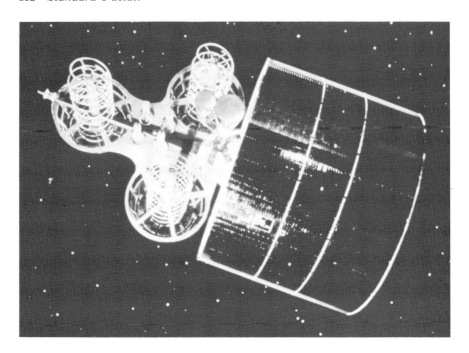

Fig. 96. Marisat satellite (courtesy INMARSAT)

Launch weight: 655 kg
Height: 3.81m
Solar array span: 2.61m
Type: Spin stabilized
Capacity: 10 2-way voice circuits
Launch vehicle: Thor Delta
Manufacturer: Hughes Aircraft Company

3.4 INMARSAT second generation space segment

The satellites comprising the first generation system will begin nearing the end of their lives during the early 1990s. The start of the design process for the second generation of satellites began in August 1983 when INMARSAT issued a call for tenders to build up to nine new satellites. A contract for the first three of this new generation of satellites, with an option for an additional six, was subsequently awarded to the British Aerospace consortium. This consortium includes companies such as the Hughes Aircraft Company, Matra Espace, and Fokker. In March 1988 INMARSAT took up part of the option and ordered a fourth INMARSAT-2 satellite to ensure sufficient communications capacity to meet the anticipated demand, and also allow for the possibility of a launch failure.

The total construction costs for the first three satellites is about £St. 130m. INMARSAT has arranged a finance package with the four main UK clearing banks in which the consortium provides the funding for the satellites and INMARSAT pays a lease charge each year of the satellites' ten year life expectancy. The satellites will, therefore, be totally owned by the consortium but operated by INMARSAT. At the end of the repayment period INMARSAT will have the option to continue to lease the satellites at a nominal charge, or to sell them, acting as the consortium's agents. If the latter option is chosen INMARSAT will retain 99% of the proceeds.

INMARSAT 2 as described above, will have some 50 times the communications capacity of the MARECS A & B2. They will be able to cover the entire allocated maritime frequencies of 15 MHz bandwidth from satellite to user, 20 MHz bandwidth from user to satellite. They will also cover 3 MHz of the aeronautical mobile satellite R band in both directions.

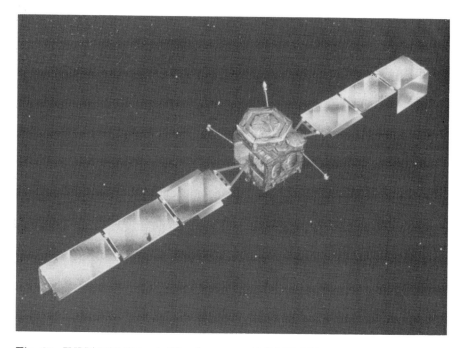

Fig. 97. INMARSAT 2 satellite (courtesy INMARSAT)

Launch weight: 1142–1271 kg
Height: 3.36m
Solar array span: 15.23m
Type: 3-axis stabilized
Capacity: 250 2-way voice circuits
Launch vehicle: probably 2 on Delta and 2 on Ariane
Manufacturer: British Aerospace Consortium
Launch dates: December 1989–June 1990

184 *Standard C detail*

3.5 INMARSAT third generation space segment

Work on developing specifications for the third generation of INMARSAT satellites began before the first satellite of the second generation was even launched! Planned to enter service from as earlier as 1994, the third generation is being specifically designed to handle large volumes of communications to very small, low-cost, light-weight user terminals, as provided by the new Standard C service. This new generation of satellites is likely to have multiple spot-beam antennas. These antennas focus on selected areas of the earth's surface instead of providing the blanket coverage of the first and second generation satellites. The main advantage of using spot-beam technology is that the number of channels provided to a given area can be tailored to demand. This leads to a more efficient use of power on the satellite, and, in combination with the store-and-forward messaging method used by the new Standard C System (see below), means that each channel is full utilised.

The combined advantages of excellent satellite power characteristics and full utilisation of communications channels, should result in a highly competitive and cost effective mode of global communications.

4. The INMARSAT Standard C system

INMARSAT's Standard C service has been available on a pre-operational basis since March 1989 in the Atlantic Ocean region (AOR) and a fully operational service will be available from November 1989 in the AOR. A fully operational world-wide service is expected from May 1990.

Standard C is a new light-weight, low-cost miniature ship earth station (SES) which is expected to greatly expand the use of maritime satellite communications by providing data communications to any size of vessel. The Standard C system utilises the existing INMARSAT space segment and has been incorporated into the design of both second and third generation space segments (INMARSAT-2 and INMARSAT-3). The new system will, however, require new coast earth stations (CES) and ship earth station installations. For a diagram of the system see Fig. 89.

Standard C is a digital communications system which provides—
1. two-way messaging and data communications on a store and forward basis
2. one-way position and data reporting
3. polling of position/data reports
4. an enhanced group call (EGC) broadcast service able to address both groups of mobiles and specifed geographic areas

Since all data is encoded into data bits, any data words, text, language or alphabet can be transmitted and the system can also handle some forms of graphics and facsimile. Data rates are 300 bits per second (bits/s) for mobiles using the first generation space segment and 600 bits/s using the second or third generation space segment.

The Standard C services are dealt with in more detail later in the chapter.

4.1 The Standard C ship earth station

Standard C ship earth stations can be designed and manufactured by any company providing it has been type approved by INMARSAT, and meets the system specifications. A ship earth station installation comprises a data circuit terminating equipment (DTE) module, and a data terminal equipment module (DTE). The data circuit terminating equipment provides the interface to the satellite network. Non-stabilised, low gain, omnidirectional antennas are used, which makes for a much simpler and cheaper construction than the gyro stabilised platform required for Standard A ship installations (see above).

Fig. 98. Standard C antenna (courtesy INMARSAT)

However, such omnidirectional antennas are prone to masking, i.e. where the line-of-site to the satellite is obscured by an obstruction. This is the same problem that occurs with GPS antennas, and the chapter on antenna location (Chapter 2, section 2.3.7) applies equally to locating Standard C antennas. Low gain omnidirectional antennas are also prone to multipath propagation, i.e. where a reflected signal from the satellite is confused with the true signal. This problem has

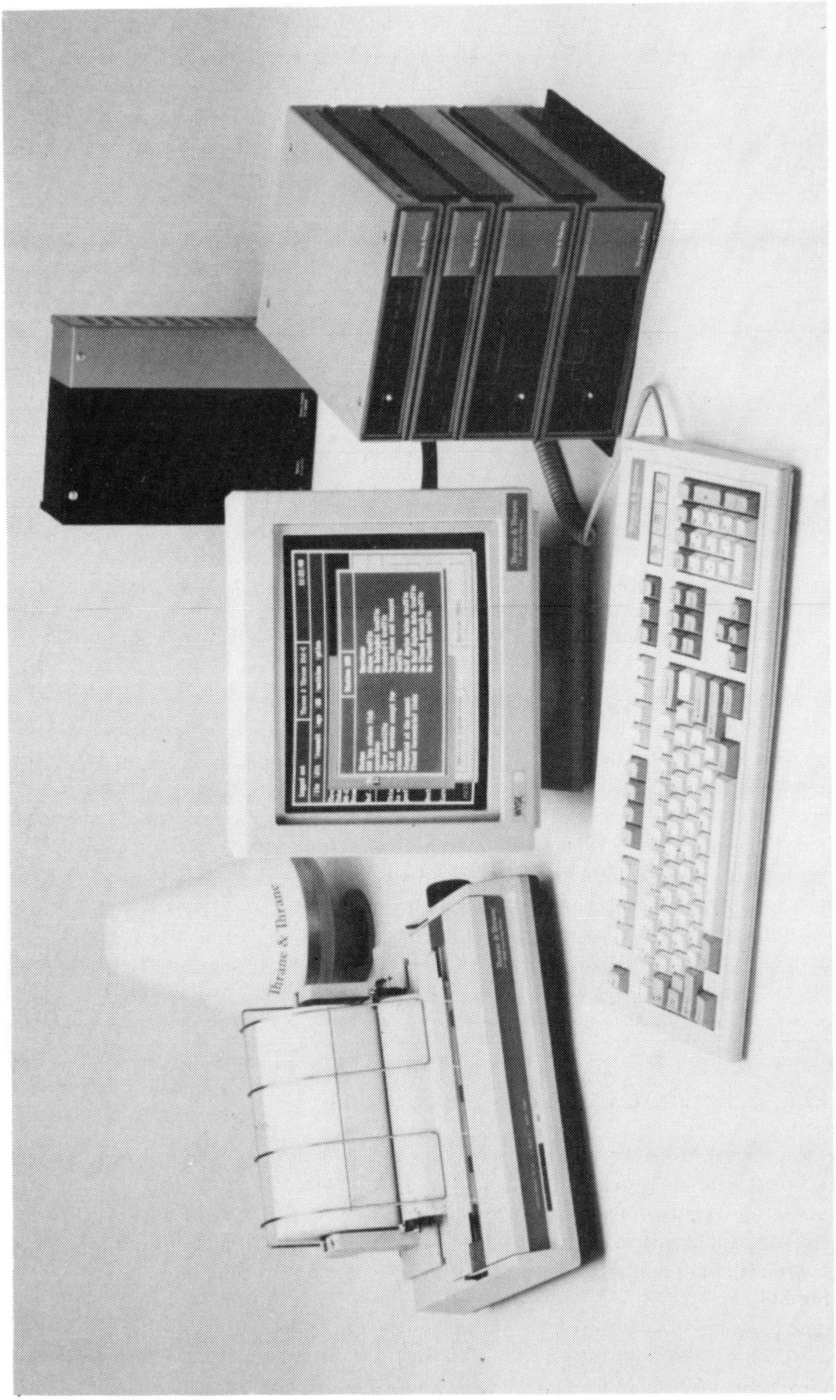

Fig. 99. The Thrane and Thrane terminal (courtesy INMARSAT)

been largely overcome by carefully choosing the message modulation and coding scheme to minimise the effects. Antenna location (above) is also important in reducing the occurrence of multipath. The data terminal equipment may range from a simple keyboard entry and display device, to a ship-board computer capable of preparing and displaying messages, as well as monitoring/ controlling various shipboard devices and functions.

The Standard C terminal must be able to tune in 5 kHz (0.005 MHz) increments throughout the 1530–1545 MHz and 1631.5–1645.5 MHz wavebands.

Three classes of DTE are envisaged—
Class one: message transfer only
Class two: either message transfer or enhanced group call at a given time
Class three: simultaneous message transfer and enhanced group call.

In January 1989, in a radical new departure for INMARSAT, the assembly formally gave approval to empower the organisation to provide land mobile satellite communications in addition to maritime and aeronautical services. The predicted number of land mobile/marine users is given below.

Table 14. Standard C terminal forecast (in thousands)

Year	1990	1991	1992	1993	1994	1995	1996	1997	1998	1999	2000	2005
Land	3	20	50	83	112	135	205	218	291	302	316	335
Marine	2	5	10	15	20	26	30	32	35	37	39	41
Total	5	25	60	98	132	162	235	250	326	339	355	376

Source: INMARSAT Standard C Newsletter No.5, April 1989

The first production Standard C terminal was available from Thrane & Thrane of Denmark in early 1989. The original units were designed to operate with the pre-operational system and were used in both maritime and land based trials during the same year. Other terminal manufacturers in at the beginning of Standard C include JRC, Furuno and Toshiba in Japan, SNEC in France and EB Nera of Norway. Overall, a large and aggressive market is expected to develop, which should serve to drive terminal prices to within the price range of many potential land and marine users. The initial prices of circa $US 10,000 should drop quickly as the market expands rapidly during the first few years of full operation and prices of $US 4–5000 should not be unrealistic.

4.2 The Standard C coast earth stations

As with Standard A, the Standard C coast earth stations are the gateway between the terrestrial network and the INMARSAT system. Telex and EGC message processing facilities are mandatory at the CES, other communications interfaces being at the discretion of the CES Operator. Several INMARSAT signatories have declared plans for building Standard C coast earth stations, and are listed in the table below.

The Standard C system also requires three new network co-ordination stations (NCS). These are to be situated at Goonhilly (UK) for the Atlantic Ocean region, Thermoplyae (Greece) for the Indian Ocean region, and Singapore for the Pacific

188 Standard C detail

Table 15. Proposed Standard C coast earth stations

Country	Location	Coverage region	Operational status
Norway	Eik	IOR	October 1989
UK	Goonhilly	AOR	November 1989
France		AOR	March 1990
USA	Southbury	AOR	April 1990
	Santa Paula	POR	May 1990
China	Beijing	IOR	May 1990
	Beijing	POR	June 1990
Denmark		AOR	August 1990
Australia		POR	November 1990
W. Germany		AOR	November 1990
Greece	Thermoplyae	IOR	December 1990
USSR		AOR	December 1990
		IOR	January 1991
		POR	February 1991
Italy	Fucino	AOR	January 1991
Singapore	Singapore	POR	January 1991

Source: INMARSAT Standard C Newsletter No.5, April 1989

Ocean region. These NCS serve the same purpose as the Standard A NCS described earlier. Also, as in Standard A, signals are transmitted to the satellite at 6 GHz and received at 4 GHz, the L-band frequencies being used to transmit (1.6 GHz) and receive (1.5 GHz) network control information.

4.3 The Standard C space segment

The various generations of INMARSAT satellites which comprise the space segment have all been discussed earlier in the chapter. We will, therefore, use this section for a more detailed description of the Standard C communications channels. There are six different channels in the Standard C system—
1. NCS common channel
2. CES TDM channel
3. SES Signalling channel
4. SES Messaging channel
5. Inter-station channel
6. Inter-region channel

4.4 The network co-ordination station (NCS) common channel

The NCS common channel is the centralised resource of the system which carries both Standard C signalling and Enhanced Group Call (EGC) messages. The NCS of an Ocean region may transmit one or more NCS common channels, each of which is permanently assigned to that NCS. For future generation satellites incorporating spot-beams, at least one NCS common channel will be transmitted in each spot beam. Each channel is based on fixed length frames of 10,368 symbols transmitted at 1200 symbols/sec, giving a time frame of 8.64 seconds and precisely 10,000 frames per day. Each frame carries a 639 byte information field, which is split into consecutive packets. The first packet in each frame is always the bulletin board packet, which contains the static operational parameters of that NCS common channel. This packet is followed by one or more signalling channel descriptor packets which describe the SES usage of the shore-to-ship CES TDMs, and the remainder of the packets are available for messaging and signalling.

To achieve low packet error rates, the signal goes through various procedures for scrambling, encoding and interleaving, which lead to an effective data rate of 600 bit/s. Access to the channel is on a priority basis, with a first-come-first-served system for packets with the same priority. There are three levels of priority for packets, which are:
1. Standard C call anouncement, EGC distress messages and distress alert acknowledgement
2. Standard C signalling
3. Other EGC messages

4.5 The coast earth station (CES) TDM channel

Each CES is able to transmit one or more TDM channels, which are assigned on demand by the Network Control Station and used when communicating with a Ship Earth Station. The channel structure is identical to that of the NCS common channel. Once again access to the channel is on a priority basis, with the priority levels being:
1. Distress packets
2. Logical channel assignments
3. Other protocol packets
4. Messages.

A unique logical channel number is assigned by the CES for each separate SES/CES connection. By providing a unique reference to an ongoing transfer they reduce protocol overheads and can only be re-used when there is no danger that it will produce ambiguities.

4.6 Ship earth station (SES) signalling channel

The CES TDM provides the forward link for each CES/SES connection and the SES signalling channel provides the return link. The SES signalling channel is based on a frame length of 8.64 seconds (the same as NCS common channel and CES TDM), which is divided into 14 slots for first generation and 28 slots for second generation satellites. The transmission rate for a burst within a slot is 600 symbols/s and 1200 symbols/s respectively. Access to the SES signalling channel can be by a reserved or unreserved method. For reserved access the slot to be used is pre-allocated by the CES. For unreserved access a random access method is employed (slotted ALOHA). The access protocol is only required for the first packet of any transmission, access for subsequent packets being guaranteed.

Since the ship earth station cannot monitor their own transmissions through the spacecraft, collision detection, e.g. where packet collision is caused by two SES

190 *Standard C detail*

trying to gain access on the same slot, is performed at the coast earth station. The result of the SES transmission as seen by the CES is returned via the CES TDM, which forms the basis of the re-transmission process. This channel is also used for the position reporting service, where small packets of data (up to 32 bytes) are transmitted over the link to a closed user group. This is a highly efficient method of sending short regular data reports without the need to resort to the CES messaging channel (see below). As there is no acknowledgement from the CES if the data received is OK (ARQ), the messages are deliberately restricted to minimise the possibility of undetected errors occurring.

4.7 SES messaging channel

An SES signalling channel is used during the call setup phase of a transfer, but the message itself is sent on a SES message channel allocated by the CES. Access to the SES messaging channel is therefore controlled by the CES. Each SES waiting to transmit a message is instructed over the CES TDM what time its transmission may start. Once assigned a start time the SES will transmit the complete message without interruption. The structure of the SES message channel is very similar to the NCS common channel and CES TDM channel. The essential differences are:
1. The SES message is quasi-continuous and therefore a pre-amble is added before transmission to aid acquisition
2. The frame length is variable between messages
3. The transmission rate is 600 symbols/s for first generation satellites and 1200 symbols/s for second generation

4.8 Inter-station channels

There is a separate dedicated channel between each CES and its associated NCS. EGC messages and anouncements generated by a coast earth station are transmitted to the NCS over the dedicated link, and subsequently broadcast over the NCS common channel. The inter-station channels are also used by the NCS to allocate CES TDM channels and ensure synchronisation of access to SES.

Inter-region channels

Each NCS is linked to the other NCS by an inter-region link channel. This link is primarily used to inform each NCS of which SES have registered in which satellite region.

5. Standard C services

As mentioned previously, there are four Standard C services provided. These are:
1. Two-way messaging and data communications on a store and forward basis
2. One-way position and data reporting

3. Polling
4. An enhanced group call (EGC) broadcast service able to address both groups of mobiles and specifed geographic areas

Before an SES can make use of any of these services, however, it must first be registered within an ocean region. Initially a SES must synchronize with an NCS common channel, and then send a login request to the NCS using the unreserved access protocol. The NCS responds to a login request with a login acknowledgement packet to the SES and a registration message to all the CES in that ocean region. The login acknowledgement packet contains network configuration information, which is essentially a list of CESs with their TDM frequencies and the services they offer for that ocean region. As part of the NCS system management task the login acknowledgement may also instruct the SES to tune into a different NCS common channel for that ocean region. The registration message is used by the CES in that ocean region to update their active ships' list, which in turn determines those ships for which they may accept calls. The NCS also informs the other NCS of the SES login so they may update their respective SES data bases.

5.1 Store and forward data and messaging

The Standard C system is capable of providing data and message communications from a mobile to a fixed communications centre such as an office, and vice versa. The communications take place on a store-and-forward basis, with the store-and-forward messages with being located at the coast earth station. For example, communications from an SES are sent in packets over the satellite link to the designated CES, where they are assembled into a complete message and then sent on to the ultimate address via the terrestrial network. In this case the store-and-forward mechanism can be thought of as three distinct message transfer processes:
1. DTE to DCE at the SES
2. SES to CES via the satellite
3. CES to terrestrial network

The store-and-forward method means efficient loading of the satellites (lower end user costs) and the digital packet structure means that a message can be anything from nautical chart corrections to weather maps or multi-lingual text messages. Full ARQ is provided to ensure error free reception of messages, and the originator is informed if the system is unable to deliver the message.

5.2 Shore-to-ship message transfer

A shore-to-ship message transfer consists of the following stages;
1. shore based originator connects to a CES over the terrestrial network,
2. message accepted by the CES,
3. call anouncement to the SES over the NCS common channel,
4. establishment of a logical channel between CES and SES,
5. message transfer over the CES TDM,
6. CES clears the CES TDM,
7. CES informs the shore based message originator of a successful message transfer.

When a shore based operator wishes to send a message to an SES he first contacts the CES over the terrestrial network and informs him of the SES ITU number. The CES then checks if that SES is, firstly, allowed to accept calls, and, secondly, is in the ocean region. If the SES is acceptable the CES then gives the go-ahead to the originator, who sends the complete message for transmission to the SES. After the complete message has been received and stored at the CES, the CES requests the NCS to announce the call over the NCS common channel. The NCS in turn checks the status of the SES. In this case the SES will be in one of three states:
1. Not in the ocean region or non-operational
2. In the ocean region and idle
3. In the ocean region and busy

This state is reported back to the CES over the interstation link channel. If the SES is busy, the NCS stores the anouncement and sends it when the SES becomes idle. If the SES is idle, the NCS transmits the announcement, along with details of the logical channel to be established between the CES and SES. The SES tunes and synchronises to the given CES TDM, then sends an assignment response over the SES signalling channel to the CES, using the unreserved access protocol. The logical channel is now established between CES and SES and the CES informs the NCS to place the SES on the busy list. The receipt of the assignment response also indicates that message transfer can now commence from the CES to the SES over the CES TDM.

The message is divided into uniquely identified packets and transmitted from the CES over the CES TDM to the SES. The CES then reserves an acknowledgement slot on the SES signalling channel, asks the SES to acknowledge that the message has been received and tells him about the reserved slot. The SES subsequently informs the CES of which packets need to be re-transmitted and when the message has been successfully received. When the CES finally receives an OK message from the SES, he sends a clear message to the SES who retunes to the NCS common channel and informs the NCS that the SES has returned to the idle state. The final task of the CES is then to inform the shore based message originator that the message has been successfully transferred. This only happens if originally requested by the message originator.

5.3 Ship-to-shore message transfer

A ship to shore message consists of a number of distinct phases, these are
1. The call request
2. Establishing the logical channel
3. Transmitting the message
4. Message acknowledgement by the CES
5. Call clearing by the CES.

Once an SES has formatted a ship-to-shore message, he uses the network configuration information obtained during ocean registration (see above) to tune into the appropriate CES TDM. After synchronising with the TDM the SES then sends an assignment request on an SES signalling channel associated with that TDM. If the CES does not serve the required final destination of the message or does not provide the required service, it will inform the SES that its assignment request has failed. If the CES is too busy to service the request it tells the SES to retune to the NCS common channel. When the CES becomes available it will then send an announcement to the SES indicating that it may now try again to establish a connection. When available, the CES, in conjunction with the NCS, goes through an assignment procedure, similar to the one described above for shore-to-ship calls, which establishes the logical channel between SES and CES. Part of this procedure involves determining the number and size of data packets within which the message will be contained and informing the NCS that the SES is now busy.

Once the logical channel is established, the SES transmits the agreed number of packets sequentially on the assigned SES message channel and starting at the indicated slot time. Upon completion of the agreed number of packets the CES informs the SES of which packets to retransmit, or, if a high proportion of the packets has been corrupted, to retransmit the complete message. When all the packets have been successfully received at the CES, it goes through its clearing procedure, informing the NCS that the SES is no longer busy. The SES retunes to the NCS common channel.

5.4 One-way position and data reporting

This section describes the Standard C Position and Data Reporting Service, which will be available for both land and maritime users. This service allows any mobile to send position reports or data reports to a designated base station (usually the fleet headquarters) from anywhere on the globe.

The position of the SES is determined by means of an onboard navigational system such as GPS, Omega, Loran C or Decca. Obviously in the context of this book and to achieve a truly global position reporting service, we are primarily concerned with the application of GPS with the Standard C position reporting service. Such applications are dealt with in Chapter 4, so here we will concentrate on how the service works.

The Position and Data reports are one-way (SES to base station), and are transmitted via the SES signalling channel. They are deliberately restricted to a maximum of 3 packets each of 15 bytes and no ARQ, or acknowledgement that the message has been received, is provided. It is a very efficient means of sending short amounts of data and, with the price per message expected to cost less than approximately 10 cents, should provide an attractive service to many users. There are two ways of using the service, referred to as, firstly, unreserved access for infrequent usage, where the SES must request resources before commencement of a report, and, secondly, reserved access for frequent repetitive usage where the users terminals are be pre-programmed to respond to a poll command (not to be confused with polling below).

For unreserved access the CES responds to an assignment request from the SES using random access by allocating a slot logical channel whose attributes include:
1. A closed network identity
2. TDM satellite frequency code
3. SES signalling channel satellite frequency code
4. Starting frame number

194 *Standard C detail*

   5. Slot number
   6. Number of packets per report (maximum 3)

Unreserved access is ideal for users who only require reports when their mobiles are undergoing particular activities, e.g. when the truck is on the highway rather than parked, or when the mobile is to report under exceptional circumstances only, e.g. cargo or engine temperatures exceed preset limits. The reserved, or pre-assigned, data reporting service is intended for users who need to gather data from SESs on a regular basis. In this case a slot logical channel is allocated which provides the SES with one or more signalling channel slots on a fixed interval basis. Reserved access, therefore, must be pre-arranged between a CES and the operator of a given group of mobiles and each group is uniquely identified by a closed network identification number (CNID).

The pre-arranged nature of this service means a more efficient use of the satellite channel capacity, as compared to the random access protocol required by the unreserved service. Because the packets require no preamble, they can also contain more user defined data than the unreserved access. In this case the slot logical channels are pre-programmed into the SESs of a given user and initiated by a group poll. The pre-programmed parameters including:
   1. Closed network identity (used by the group poll)
   2. Starting frame number
   3. Slot number
   4. Number of packets per report (maximum 3)
   5. Reporting interval (how often to report)
   6. Assignment duration

The channel and frequency allocations to be used for the response are contained in the group poll message. The report interval is given in terms of frames, there being 10,000 frames in 24 hours (1 frame = 8.64 seconds). The reporting interval can be set to between 10 and 63,000 frames (approximately 6.3 days). The assignment duration can be set from one to 63,000 reports, or unlimited if requested.

5.5 Report format

Each report consists of a minimum of one 15 byte data packet and a maximum of three 15 byte data packets. Of the 15 bytes in each packet, 3 are always taken up by protocol information, leaving a maximum of 12 × 3 or 36 bytes for data. If the report is in the unreserved format a further 4 bytes of the first packet is taken up by addressing information, reducing the maximum data size to 32 bytes. There are two styles of position report, the maritime position report, and the land mobile position report. The maritime position report contains exactly 12 bytes of predetermined data, which includes position (latitude and longtitude), macro encoded message (MEM), course, speed and ETA of the vessel.

The Land Mobile Position report also contains exactly 12 bytes of predetermined data. In this case position (latitude and longtitude), MEM, ETA, and mileage.In the reserved format a complete report can be obtained in one packet (12 + 3 = 15), leaving the other two packets (24 bytes) free for user defined data. In the unreserved format a one packet report will not include ETA (maritime) or mileage (land), and all three packets leave 20 bytes free for user defined data, e.g. cargo temperatures, engine revolutions etc. Position report sub-types are also supported, allowing the reports to be configured in existing, or planned, national and international formats such as AMVER (American version), JAPREP (Japanese report), AUSREP (Australian) or a maritime weather report format.

5.6 Polling

The polling facility can be used for one of the following purposes:
1. Initiating data reports from an SES or specified group of SESs
2. Sending a text or data message to an SES or group of SESs
3. Remotely programming an SES or group of SESs with the parameters required for the Reserved Data Reporting Protocol (see above)

Closed network addressing is used to identify the specified group of SESs, and individual devices on the SESs can be activated by specifying their sub-address within the polling message. Groups of SESs can be specified in one of three ways:
1. a list of one or more individual SES identity numbers
2. a Closed Network Identity
3. a Closed Network Identity within a given geographical area

These are usually referred to as individual, group and area polling.

With individual polling a separate request is sent to each SES over the NCS common channel, whereas with group and area polling only one message is broadcast. Polling requests can contain up to 256 characters of text message, which will be received by all the addressed SESs. They can also contain instructions on how the SESs should respond, what data they should return, and in what format the data should be contained. For example, data could be transmitted back to shore using either the position reporting or standard message services.

5.7 Enhanced group calls

The enhanced group call (EGC) system allows INMARSAT to provide a global one-way shore-to-ship message broadcast service to predetermined groups of mobiles in both fixed and variable geographical areas. EGC messages can originate from any authorised subscriber, and are sent via terrestrial links to a coast earth station (CES). The message is processed at the CES and then forwarded to a network co-ordination station (NCS) for broadcast over an NCS common channel. The message will only be received by vessels which have been programmed for that subscriber.

Vessel receivers are addressed on the basis of:
1. Individual unique ID
2. Group Ids
3. Pre-assigned geographical areas
4. Temporary geographical areas

To receive geographically addressed broadcasts a vessel must have knowledge of its own position. However in every other way this service is totally compatible with Standard C. Two types of service are available, SafetyNET for the broadcast of global, regional or local maritime safety information, and FleetNET for

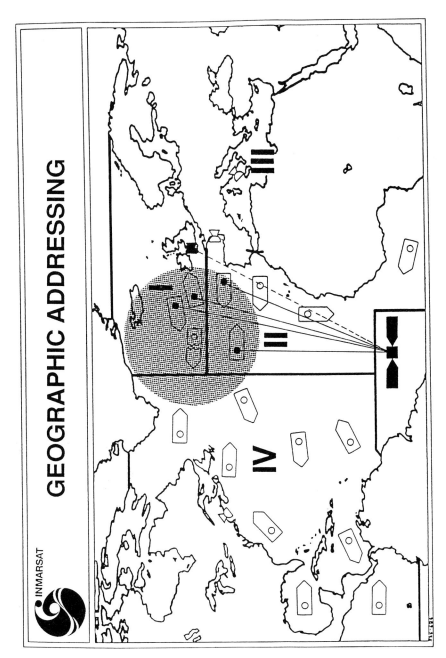

Fig. 100. Geographic addressing (courtesy INMARSAT)

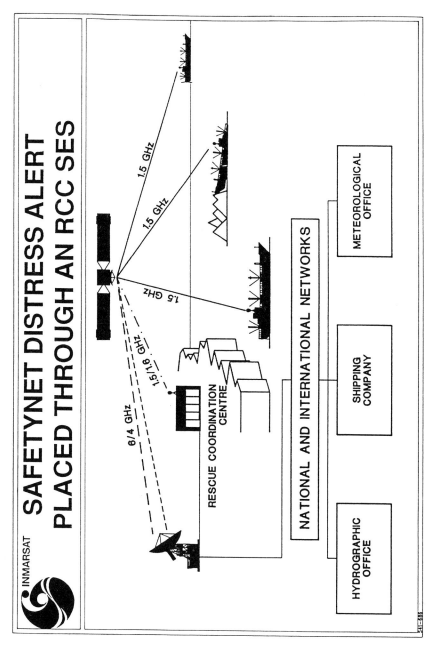

Fig. 101. SafetyNet a distress alert placed through an RCC SES (courtesy INMARSAT)

commercial users. The SafetyNET service is designed to meet the requirements of authorities and administrations, who need to broadcast maritime safety information, such as NAVAREA and navigation warnings, meteorological warnings and forecasts, shore-to-ship distress alerts, and other urgent information. The FleetNET service can be accessed by a national fleet, company fleet or commercial subscription service such as news or stock market information. The messages are secure and can only be received by pre-programmed receivers.

For a diagram showing SafetyNet call for NAVAREA II see Fig. 54. For a diagram showing the FleetNet call for NAVAREA II see Fig. 55.

5.8 Standard C and the global maritime distress and safety system

The Standard C system, and the SafetyNET service in particular, play an important role in the future global maritime distress and safety system (GMDSS). All of the medium and long range communications functions identified by the International Maritime Organisation (IMO), as being essential capabilities for ships in the GDMSS can be met by the Standard C/SafetyNET combination (Safety at Sea, pp12–14, May 1989).

Standard C has been accepted by the IMO as primary communication equipment for all ships of 300 gross tonnage and over, operating in area A3 (deep sea) of the GDMSS. In addition to covering the mid-ocean regions, the EGC system can also be adapted to provide an automated 518 KHz NAVTEX compatible service for coastal areas, where countries may not find it economical to install a terrestrial based system.

Index

Active Glonass satellites, 127

2690 BT ARPA, Sealink ferry Horsa, 82
Buoy monitoring, 96–98
Buoy movements, 96–98

Caesium clock, 138
Cartesian reference frames, 161–162
Chayka, 6–9
  coverage, 7
Coastal DGPS service, 70
Coastal navigation, 69–90
  conventional aids, 77–80
  conventional electronic positioning services, 71–74
    accountability, 74
    operator costs, 74
    precise positioning in North Sea, 72
    user acesss, 71–73
    user cost, 73
  daymarks, 78
  definition, 69
  light sources, 78
  navigation equipment on merchant vessels, 80–83
  equipment rationalization, 81
  physical factors, 70
  Racon, 80
  radar aids, 78
  radio aids, 78
  sound sources, 78
  technological factors, 70–71
  traffic management, 74–77
  traffic separation schemes, 74–77
    specification, 76–77
Coastguards, GPS for, 87–89
Codeless GPS, 148–151
  carrier aided filtering, 149–150
  phase differencing, 150–151
Co-sited MF transmitter and lighthouse, 79

Decca, 9
  coverage, 8
Diffcell/Microtel, 67
Differential GPS, 45–68
  common view (block shift), correction

Differential GPS—*cont.*
  technique, 48
  concept, 45–49
  data links, 59–68. *See also* Data links
  differential error budget, 47
  improvement of performance, and, 45
  ionospheric component, 47
  integrity monitoring, 58–59
  limiting factors, 46
  pseudolites, 52—53
  pseudo-range comparisons, 51
  pseudo-range corrections, 49–52
    advantages, 49–52
    disadvantages, 49–52
  pseudo-range error sources, 50
  quality control, 58–59
  Radio Technical Committee for Marine Services, 55–58
  selective availability, 53–55
    plot, 54
  system design, 46–48
  techniques of correction, 49–52
  tropospheric delay, 47
Data links, 59–68
  compromise, 60–61
  differential options, 61–68
    long wave/low frequency 30 kHz to 300 kHz, 61
  error correction, 60
  error detection, 60
  existing kHz differential coverage, 63
  high frequency, 3 mHz to 25 MHz, 66
  medium frequency, 300 kHz to 3 MHz, 64–66
  medium frequency d'Hevential station, 65
  satellite dimension, 66–68
  100 kHz transmitter, 62
  UHF, 66
  VHF, 66
Dredging operations, 96
Dredging system, 97

EEZ management, GPS for, 90
Einstein, GPS clock, and, 141
Electronic fleet management systems, 108, 115
Engine Monitoring and Control Systems (EMCS), 29

199

Fishermen, GPS for, 84
Fishing vessel, electronic, 85
Fleetnet call, 111

GEO/HIO mix, 13
Geostar, 12
   coverage, 11
Global positioning, 1–13
   concept, 2–3
   global ship, 2
Global positioning system. *See* GPS
Global ship, 2
Glonass, 3, 4
GPS
   alternative applications, 83–90
   applications, 69–118
   codeless. *See* Codeless GPS
   differential, 45–68. *See also* Differential GPS
   electronic bridge, and, 27–31
   first systems, 14–16
   implications, 69–118
   integrity monitoring, 58–59
   navigator's choice, 4
   operator, and, 31
   quality control, 58–59
   satellite, 3
   satellite positioning system configuration, 120
   ship, and, 27–44
   system design and implementation, 119–138
GPS clock, 138–141
   Einstein, and, 141
   fundamental clock, 140
   system time, 140–141
   oscillator performance comparison, 139
GPS detail, 119–170
GPS frequencies, 141–148
   Binary biphase modulation, 144–145
   C/A code, 145–147
   carriers, 141–143
   codes, 143–145, 146
      Navstar, 144–145
   Glonass, 143
   GPS signal in quadrature, 145
   Navstar, 141–143
   P code, 147–148
   spread spectrum, 143–145
   Y code, 148
GPS positioning quality, 163–170
   completed constellation, 167–168
   confidence levels, 168–170
      accuracies, 169
   geometry outages, 168
GPS receiver, 31–44
   composite Navstar and Glonass, 35–36
   dual channel, 32
   fast sequencing, 32–34
   hybrid, 34–36

GPS receiver—*cont.*
   integrated GPS/Loran C receiver, 36
   multiplexing, 34
   parallel/multi-channel receivers, 31–32
   practical guide to purchase, 36–44
      antenna installations and cabling, 41–43
      differential features, 43
      generic receiver menu design, 39
      hardware integration, 40
      hardware interfacing, 39–40
      navigation features, 37
      operating modes, 37–38
      power supplies, 41
      service agreements, 43–44
      software upgrades, 43–44
      user interface, 38–39
   slow sequencing, 32–34
   types, 31–34
GPS receiver design, 131–132
GPS Shipmate RS 5310 A multiplexing rx, 33
GPS signal spectrum, 132
Ground/control segment, 127–130
   civilian GPS information centre, 129
   Glonass, 129

Hydrographic surveys, 91–96
   automated system, 94
   computer hardware, 96
   fully automated surveys, 95
   low technology, 91–92
   semi-automated survey package, 93
   semi-automated surveys, 92
   software complexity, 95
   software language, 95

INMARSAT, 171–197
   CES parabolic antenna, 179
   current (first generation) space segment, 178–182
      frequency allocation, 181
      satellites, 179
   global coverage, 173
   Inmarsat 2 satellite, 183
   Intelstat V satellite, 181
   Marecs satellite, 180
   Marisat satellite, 182
   organization, 171–173
   second generation space segment, 182–183
   Standard A service, 174–184. *See also* Standard A service system, 172
   third generation space segment, 184
INMARSAT position reporting and surveillance service, 112
INMARSAT Standard C system, 109–115, 184–197. *See also* Standard C system
   basic position reporting system, 112
   enhanced group call facility, 109–111
   position reporting and display system, 112–114

INMARSAT Standard C system—*cont.*
  communications, 114
  position display and logging, 114
  position report database, 114
  reports and data export, 114
  position reporting service, 113
Integrated bridge, 28
Integrated fleet management, 115–117
  fleet network, 116
Integrated Navigation Systems (INS), 29
Integrated ship, 27–31
  fuel efficiency, 29

Kepler, 154
Keplerian parameters, 154

Loran C, 6–9
  coverage, 7
LSR 4000 Live Situation Report, 29

Magnavox MX 4400 GPS positioning and
    navigation system, 42
Merchant vessels,
  GPS for, 83–84
  navigation equipment on, 80–83
Micro-fix beacon, 106
MNS 2000 G display, 35

Nav Graphic II, 76
Navigation, satellites, and, 1–26
NAVSAT, 12–13
Navstar GPS, 3, 4, 17–26
  accuracies, 23–26
  C/A Code modulations, 19
  calculation of position, 23
  current configuration, 18
  differential scatter plot, 26
  Dilution of Precision, 23–25
  English Channel coverage, 18
  enhanced accuracy levels, 25
  geometry effects, 24
  navigating with, 19–23
  orbit, 17
  P Code modulations, 19
  position aiding, and, 22
  pseudo-range measurements, 20–22
  single-receiver accuracies, 25
  system design, 17–18
Navstar GPS Block 1 in-plane constellation,
    124
Navstar GPS CGIC interface, 130
Navstar GPS satellite, 123

Officer of the Watch, functions and duties,
    27–28
Oilmen, GPS for, 87
Omega, 4–5
  station locations, 5
Operational ground control system Navstar
    GPS, 128

Pilotage, 98–101
  electronic chart display system, 99–101
  display specification, 99–101
  operating specifications, 99
  physical specification, 99
  software specification, 99
Policemen, GPS for, 87–89
Port DGPS service, 102
Port pilotage system, 100
Port positioning, 90–107
  availability, 107
    conventional, 107
    GPS, 107
  buoy monitoring, 96–98
  buoy movements, 96–98
  conventional services, 91–101
  dedicated beacons, 90–91
  demand restrictions, 107
    conventional, 107
    GPS, 107
  dredging operations, 96
  geography, 105–106
    conventional, 105
    GPS, 105–106
  GPS service, 101–107
    accurate service, 104–107
    communication protocols, 101
    compatibility, 101–103
    co-ordinate systems, 101
    levels of service, 103–107
    physical characteristics, 103
    precision service, 103
  hydrographic surveys, 91–96. *See also*
      Hydrographic surveys
  operational considerations, 106
    conventional, 106
    GPS, 106
  pilotage, 98–101
  Port DGPS service, 102
  reliability, 107
    conventional, 107
    GPS, 107
  standard service, 107
  vessel navigation, 98–101
Position and data reporting, 108–117
Pseudo-range, meaning, 152
Pseudo-ranging for position, 152–163
  Cartesian and geodectic frameworks, 16
  Cartesian reference frames, 161–162
  computation to position, 154, 155
  ellipsoid, 160–161
    associated figures of earth, 160
    simple ellipse, 161
  GPS ranging principle, 155
  position aiding, 156–159
    clock aiding, 159
    height aiding, 156–159
  position reference frame, 159–163
  pseudo-range, 152, 153

Pseudo-ranging for position—*cont.*
  range to position, 156
  satellite's position, 152, 154
  WGS 84 Geoid-spheroid separation, 158
  WGS 84 spheroid, 162–163
Pseudolites, 52–53

Quartz clock, 138

Racal Decca MNS 2000, 30
Radio Technical Committee for Marine Service, 55–58
  data rate, 57–58
  message format, 55–58
    Type 1, 56–57
    types, 57
Real-time quality control, 163–167
  dilution of precision, 164–166
    diamond of errors, 164
    user equivalent range error (UERE), and, 166–167
Receiver design, 134
Regional navigation system, 4–9
  satellite alternatives, 9–13
  proposed systems, 12–13
Rubidium clock, 138

Safety Net, distress alert, 198
Safety Net call Navarea II, 110
SARSAT-COSPAS, 89
Satellite constellation Bird-cage, 121
Satellite positioning system configuration, 120
Satellites, navigation, and, 1–26
Sealink ferry *Hengist*, 30
Sealink ferry *Horsa*, 82
Seismic vessel, towing sound source, 86
Shipmate RS 2500 colour plotter, 41
Signals, 138–151
Space segment, 120–127
  orbit design, 121–122
    Glonass, 122
    Navstar GPS, 121–122
  satellites, 122–127
    Block 1 and planned Block 2 positions, 126
    Glonass, 127
    Navstar GPS, 122–126
    Navstar numbering schemes, 125
Standard A service, 174–184
  below-deck installation, 176
  coast earth stations, 177–178
    table, 178
  radome installation, 175
  ship earth station, 175–177

Standard C system, 184–197
  antenna, 185
  coast earth stations, 187–188
    TDM channel, 189
  enhanced group calls, 195
  geographic addressing, 196
  global maritime distress and safety system, and, 197
  inter-station channels, 190
  inter-region channels, 190
  network co-ordination station (NCS), common channel, 188–189
  one-way position and data reporting, 193–194
  polling, 195
  report format, 194–195
  Safety Net distress alert, 198
  services, 190–198
  SES messaging channel, 190
  ship earth station, 185–187
  signalling channel, 189–190
  ship-to-shore message transfer, 192–193
  shore-to-ship message transfer, 191–192
  space segment, 188
  store and forward data and messaging, 191
  terminal forecast, 187
  Thrane and Thrane terminal, 186
Star-Fix, 9–12
  system coverage, 10
System status, 136–138
  block 2 launch schedule, 136
  current and projected launch build up, 137–138

Transit (Sat Nav), 14–16
  orbit arrangement, 14
  perspective, 16
  system principle, 15

User segment, 131–136
  from measurement to position, 136
  from noise to signal, 132–133
  from numbers to code, 133–135
  from numbers to phase, 135
  from signal to numbers, 133
  GPS receiver design, 131–132

Vessel identification, 88
Vessel interception, 88–89
Vessel surveillance, 88

Yachtsmen
  GPS for, 89–90